PERIODIC TABLE OF THE ELEMENTS

PERIODIC TABLE OF THE ELEMENTS

Groups	I	II	III	IV	V
2	6.939 **3** **Li** Lithium (1 2)	9.0122 **4** **Be** Beryllium (2 2)	**5** 10.811 **B** Boron (2 3)	**6** 12.01115 **C** Carbon (2 4)	**7** 14.0067 **N** Nitrogen (2 5)
3	22.9898 **11** **Na** Sodium (1 8 2)	24.312 **12** **Mg** Magnesium (2 8 2)	**13** 26.9815 **Al** Aluminium (3 8 2)	**14** 28.086 **Si** Silicon (4 8 2)	**15** 30.9738 **P** Phosphorus (5 8 2)
4	39.102 **19** **K** Potassium (1 8 8 2)	40.08 **20** **Ca** Calcium (2 8 8 2)	44.956 **21** **Sc** Scandium (2 9 8 2)	47.90 **22** **Ti** Titanium (2 10 8 2)	50.942 **23** **V** Vanadium (2 11 8 2)
	29 63.546 **Cu** Copper (1 18 8 2)	**30** 65.37 **Zn** Zinc (2 18 8 2)	**31** 69.72 **Ga** Gallium (3 18 8 2)	**32** 72.59 **Ge** Germanium (4 18 8 2)	**33** 74.9216 **As** Arsenic (5 18 8 2)
5	85.47 **37** **Rb** Rubidium (1 8 18 8 2)	87.62 **38** **Sr** Strontium (2 8 18 8 2)	88.905 **39** **Y** Yttrium (2 9 18 8 2)	91.22 **40** **Zr** Zirconium (2 10 18 8 2)	92.906 **41** **Nb** Niobium (1 12 18 8 2)
	47 107.868 **Ag** Silver (1 18 18 8 2)	**48** 112.40 **Cd** Cadmium (2 18 18 8 2)	**49** 114.82 **In** Indium (3 18 18 8 2)	**50** 118.69 **Sn** Tin (4 18 18 8 2)	**51** 121.75 **Sb** Antimony (5 18 18 8 2)
6	132.905 **55** **Cs** Cesium (1 8 18 18 8 2)	137.34 **56** **Ba** Barium (2 8 18 18 8 2)	138.91 **57** **La** Lanthanum (2 9 18 18 8 2) * 58-71	178.49 **72** **Hf** Hafnium (2 10 32 18 8 2)	180.948 **73** **Ta** Tantalum (2 11 32 18 8 2)
	79 196.967 **Au** Gold (1 18 32 18 8 2)	**80** 200.59 **Hg** Mercury (2 18 32 18 8 2)	**81** 204.37 **Tl** Thallium (3 18 32 18 8 2)	**82** 207.19 **Pb** Lead (4 18 32 18 8 2)	**83** 208.980 **Bi** Bismuth (5 18 32 18 8 2)
7	[223] **87** **Fr** Francium	[226] **88** **Ra** Radium	[227] **89** **Ac** Actinium ** 90-103	**104** **Ku** Kurchatovium	**105**

140.12 **58** **Ce** Cerium (2 8 20 18 8 2)	140.907 **59** **Pr** Praseodymium (2 8 21 18 8 2)	144.24 **60** **Nd** Neodymium (2 8 22 18 8 2)	[145] **61** **Pm** Promethium (2 8 23 18 8 2)	150.35 **62** **Sm** Samarium (2 8 24 18 8 2)	151.96 **63** **Eu** Europium (2 8 25 18 8 2)	157.25 **64** **Gd** Gadolinium (2 8 25 19 8 2)
232.038 **90** **Th** Thorium (2 8 18 32 18 10 2)	[231] **91** **Pa** Protactinium (2 8 18 32 20 9 2)	238.03 **92** **U** Uranium (2 8 18 32 21 9 2)	[237] **93** **Np** Neptunium (2 8 18 32 22 9 2)	[242] **94** **Pu** Plutonium (2 8 18 32 23 9 2)	[243] **95** **Am** Americium (2 8 18 32 24 9 2)	[247] **96** **Cm** Curium (2 8 18 32 25 9 2)

	VII	VIII	Periods →	
	1.00797 **H** 1 Hydrogen 1	4.0026 **He** 2 Helium 2	1	**He** – s-elements
VI				**Ne** – p-elements
8 15.9994 **O** Oxygen	9 18.9984 **F** Fluorine	20.183 **Ne** 10 Neon	2	**Hg** – d-elements
16 32.064 **S** Sulphur	17 35.453 **Cl** Chlorine	39.948 **Ar** 18 Argon	3	**Td** – f-elements
51.996 24 **Cr** Chromium	54.9380 25 **Mn** Manganese	26 55.847 **Fe** Iron	27 58.9332 **Co** Cobalt	28 58.71 **Ni** Nickel
34 78.96 **Se** Selenium	35 79.904 **Br** Bromine	83.80 **Kr** 36 Krypton	4 Atomic masses and mass-numbers of longest-lived isotopes (in brackets) are International 1965 values	
95.94 42 **Mo** Molybdenum	[99] 43 **Tc** Technetium	44 101.07 **Ru** Ruthenium	45 102.905 **Rh** Rhenium	46 106.4 **Pd** Palladium
52 127.60 **Te** Tellurium	53 126.9044 **I** Iodine	131.30 **Xe** 54 Xenon	5	
183.85 74 **W** Tungsten	186.2 75 **Re** Rhenium	76 190.2 **Os** Osmium	77 192.2 **Ir** Iridium	78 195.09 **Pt** Platinum
84 [210] **Po** Polonium	85 [210] **At** Astatine	[222] **Rn** 86 Radon	6	Mass number of longest-lived isotope
106				Atomic number (number of electrons in atom) 84 [210] **Po** Number of electrons in shell
				Atomic mass (C-12 scale)

158.924 65 **Tb** Terbium	162.50 66 **Dy** Dysprosium	164.930 67 **Ho** Holmium	167.26 68 **Er** Erbium	168.934 69 **Tm** Thulium	173.04 70 **Yb** Ytterbium	174.97 71 **Lu** Lutetium
[247] 97 **Bk** Berkelium	[249] 98 **Cf** Californium	[254] 99 **Es** Einsteinium	100 **Fm** Fermium	101 **Md** Mendelevium	102 **(No)** Nobelium	103 **Lr** Lawrencium

PROBLEMS
IN GENERAL CHEMISTRY

by

N. L. GLINKA

Translated from the Russian

by

A. A. ROSINKIN

University Press of the Pacific
Honolulu, Hawaii

Problems in General Chemistry

by
N. L. Glinka

ISBN: 1-4102-2589-5

Reprinted from the 1973 edition

University Press of the Pacific
Honolulu, Hawaii
http://www.universitypressofthepacific.com

CONTENTS

FOREWORD

TO THE THIRTEENTH

RUSSIAN EDITION

During preparation of this textbook for print the material of the previous edition was thoroughly revised and the numerical data checked and reverified. Also introduced into the book were corrections made in accordance with decisions of the International Union of Pure and Applied Chemistry adopted in 1961, concerning the system of atomic weights, in which the atom of carbon-12(^{12}C) is utilized as the standard and is assigned a value of exactly 12.

Since this textbook is in a measure a supplement to my *General Chemistry*, corrections were introduced into this textbook to coordinate its material with the essential revisions made in the theory.

In this textbook, preference is given to the newly adopted International System of units, hence the divergence from designations and numerical data found in previous editions of the *General Chemistry*.

The author wishes to express his gratitude to L. V. Brovkin who prepared this edition for print.

N. L. Glinka

FOREWORD
TO THE THIRTEENTH
RUSSIAN EDITION

CHAPTER I

COMPOSITION OF COMPOUNDS

1. Composition of Chemical Compounds

The composition of a chemical compound is characterized by the ratio of masses of the elements which compose it. This ratio can be established either by building up a compound out of simple substances (synthesis) or by decomposing a complex substance into simple ones (analysis).

Example 1. While determining the composition of magnesia 6 g of magnesium were burned to form 10 g of magnesium oxide. It follows therefore that 10 g of magnesia contain 6 g of magnesium and 4 g of oxygen, in other words, magnesium and oxygen combined in the ratio of 6 : 4 or 3 : 2. This ratio expresses the composition of magnesium oxide.

Example 2. In decomposition of water with electric current 0.5 g of hydrogen and 4 g of oxygen evolved. It follows from these data that the composition of water can be expressed by the ratio 0.5 : 4 or 1 : 8. In other words, water contains one part by weight of hydrogen per eight parts by weight of oxygen.

* * * *

The composition of a compound is very often expressed in per cent. To this end 100 is divided into parts proportionate to the numbers in the ratio describing the given compound. For example, the composition of magnesium oxide can be expressed in per cent by weight as follows:

$$\text{per cent magnesium in MgO} = \frac{100 \times 3}{5} = 60$$

$$\text{per cent oxygen in MgO} = \frac{100 \times 2}{5} = 40$$

Practically, the composition of a compound is determined not only by its synthesis or analysis but also by the reactions in which the given substance participates or is obtained as the resultant product.

Example 3. When reducing 8 g of copper oxide by carbon, 2.2 g of carbon dioxide have evolved whose composition is characterized by the ratio 3 : 8 (3 weight parts of carbon per 8 weight parts of oxygen). Calculate the composition of copper oxide.

Solution. Determine by proportionate division the weight of oxygen contained in 2.2 g of carbon dioxide:

$$\frac{2.2 \times 8}{11} = 1.6 \text{ g}$$

Since this is the weight of oxygen contained in 8 g of copper oxide before the reaction, the weight of copper in 8 g of copper oxide is $8 - 1.6 = 6.4$ g.

Hence, the composition of copper oxide can be expressed by the ratio (Cu) : (O)=6.4 : 1.6 or 4 : 1.

PROBLEMS

1. In preparation of iron sulphide 2.4 g of sulphur combined with 4.2 g of iron. Express the composition of this compound by the ratio of the least whole numbers and in per cent.

2. On decomposition of 7 g of calcium bromide 5.6 g of bromine were obtained. Express the composition of calcium bromide by the ratio of whole numbers and in per cent.

3. When burned, 1.55 g of phosphorus produce 3.55 g of phosphoric anhydride. What is the quantitative composition of phosphoric anhydride?

4. On burning a certain amount of a hydrocarbon, 3.3 g of carbon dioxide and 2.7 g of water were formed. Determine the quantitative composition of the hydrocarbon (for the composition of carbon dioxide and water see Examples 2 and 3).

5. On burning 1.9 g of carbon sulphide, 3.2 g of sulphurous anhydride containing one part by weight of sulphur per one part by weight of oxygen were formed. Determine the composition of carbon sulphide and express it in per cent.

6. The composition of silver sulphide is expressed by the ratio (Ag) : (S)=24 : 4. How many grams of silver can be obtained from 124 g of this substance?

7. On burning a certain amount of hydrogen arsenide 0.66 g of arsenous acid anhydride containing 75.8 per cent of arsenic and 0.18 g of water were obtained. Express the composition of hydrogen arsenide in per cent.

8. On burning 16 g of cupric sulphide (containing 80 per cent of copper and 20 per cent of sulphur), 16 g of copper oxide and 6.4 g of sulphurous anhydride were obtained. Determine the composition of the obtained compounds and express their ratios with the least whole numbers.

9. How many grams of water will be formed on burning 20 g of a hydrocarbon containing 25 per cent of hydrogen and 75 per cent of carbon? The composition of water is expressed by the ratio (H) : (O)=1 : 8.

10. On decomposition of 50 g of calcium carbonate, 28 g of calcium oxide and 22 g of carbon dioxide were obtained. What is the composition of calcium carbonate if calcium oxide contains 5 parts by weight of calcium per 2 parts by weight of oxygen, and carbon dioxide contains 3 parts by weight of carbon per 8 parts by weight of oxygen?

2. Equivalent Weight.
The Law of Multiple Proportions

By the *equivalent weight of an element* is understood the weight of the element which will combine with eight parts by weight of oxygen (or 1 part of hydrogen) or will displace these amounts from their compounds.

Elements combine with one another in quantities proportional to their equivalent weights (the **law of equivalent weights**).

It follows from the law of equivalent weights that:

(1) numbers 8 for oxygen and 1 for hydrogen are the equivalent weights for these elements;

(2) the equivalent weight of an element can be determined from the composition of its compound with another element whose equivalent weight is known.

If an element forms several compounds with another element its equivalent weight will evidently have different magnitudes depending on a particular compound used for its calculation. These different values of the equivalent weight will however relate to each other as small whole numbers.

If two elements form several chemical compounds with one another, the weights of one of the elements corresponding to a fixed weight of the other in these compounds are in a simple integral proportion (the **law of multiple proportions**).

The concept of the equivalent weight and the law of equivalent weights hold true also for complex substances.

The *equivalent weight of a complex substance* is the weight which will react without residue with one equivalent weight of hydrogen (1 part by weight) or oxygen (8 parts by weight).

Thus, it follows that the substances react with one another in weight quantities which are proportional to their equivalent weights. Therefore, in order to determine an equivalent weight of a simple or complex compound, it will only suffice to establish the weight proportion in which a given substance reacts with any other substance whose equivalent weight is known.

The quantity of a substance in grams equal to its equivalent weight is known as the *gram-equivalent*.

Example 1. Determine the equivalent weight of magnesium if its 3 parts by weight combine with 2 parts by weight of oxygen.

Solution. According to the law of equivalent weights the quantities of the reacting magnesium and oxygen should be proportional to their equivalent weights. By designating the equivalent weight of magnesium as E_{Mg} and bearing in mind that the equivalent weight of oxygen is 8, the following proportion can be derived:

$$3:2 = E_{Mg}:8$$

whence

$$E_{Mg} = \frac{3 \times 8}{2} = 12$$

Example 2. Calcium chloride contains 36 per cent of calcium and 64 per cent of chlorine. Determine the equivalent weight of calcium if the equivalent weight of chlorine is 35.5.

Solution. The weight ratio of calcium to chlorine in calcium chloride is 36 : 64. It should be equal to the ratio of their equivalent weights. Designate the equivalent weight of calcium by E_{Ca} and make out a proportion:

$$36:64 = E_{Ca}:35.5$$

whence

$$E_{Ca} = \frac{36 \times 35.5}{64} = 20$$

Example 3. Determine the equivalent weight of phosphoric acid, if on substitution of calcium for hydrogen, 0.612 part by weight of calcium (whose equivalent weight is 20) reacts with one part by weight of the acid.

Solution. Following the model of the previous example, make out a proportion:

$$1:0.612 = E_{acid}:20$$

whence the equivalent weight of the acid is

$$E_{acid} = \frac{1 \times 20}{0.612} = 32.7$$

PROBLEMS

11. When burned, 5 g of aluminium form 9.44 g of alumina. Determine the equivalent weight of aluminium.

12. Sulphide of a metal contains 52 per cent of the metal. The equivalent weight of sulphur is 16. Determine the equivalent weight of the metal.

13. On dissolution of 3.06 g of a metal in an acid 2.8 litres of hydrogen, measured at a temperature of 0°C and pressure of 760 mm Hg, were evolved. Calculate the equivalent weight of the metal.

14. Determine the equivalent weight of a metal whose 2 g displace 1.132 g of copper from a solution of its salt. The equivalent weight of copper is 31.8.

15. 1 g of a metal combines with 1.78 g of sulphur or 8.89 g of bromine. Determine the equivalent weights of bromine and the metal knowing that the equivalent weight of sulphur is 16.

16. 1.6 g of calcium and 2.615 g of zinc displace equal quantities of hydrogen from an acid. The equivalent weight of calcium is 20. Determine the equivalent weight of zinc.

17. A certain quantity of a metal whose equivalent weight is 28 displaces from an acid 700 ml of hydrogen measured at standard conditions *. Determine the mass of the metal.

18. A quantity of a metal combines with 0.02 g of oxygen and 3.173 g of a halogen. Determine the equivalent weight of the halogen.

19. The mass of 1 litre of oxygen is 1.4 g. How many litres of oxygen are required to burn 21 g of magnesium whose equivalent weight is 12?

20. To reduce 1.8 g of metal oxide, 833 ml of hydrogen were spent (as measured at standard conditions). Calculate the equivalent weight of the metal and its oxide (see the footnote to Problem 17).

21. The oxide of a metal contains 28.57 per cent of oxygen, and the fluoride of the same metal contains 48.72 per cent of fluorine. Calculate the equivalent weight of fluorine.

22. To dissolve 16.86 g of metal, 14.7 g of sulphuric acid whose equivalent weight is 49 were spent. Determine the equivalent weight of the metal and the volume of hydrogen evolved during its dissolution.

23. 1.355 g of ferric chloride reacts with 1 g of sodium hydroxide without residue. The equivalent weight of NaOH is 40. What is the equivalent weight of ferric chloride?

24. In neutralization of an acid 1 g of sodium hydroxide reacts with 1.125 g of acid. The equivalent weight of sodium hydroxide is 40. Determine the equivalent weight of the acid.

25. Arsenic produces two oxides, one of which contains 65.2 per cent and the other 75.8 per cent of arsenic. Determine the equivalent weights of arsenic. How do these equivalents relate to each other?

26. Tin forms two oxides, containing 78.8 and 88.12 per cent of tin. Determine the equivalent weights of tin from the composition of these oxides and find the ratio of these equivalents.

---

* The mass of one litre of hydrogen in these conditions is 0.09 g.

CHAPTER II

THE GAS LAWS.
DETERMINATION OF MOLECULAR WEIGHTS
(MOLECULAR MASSES) OF GASES
AND VAPOURS

1. Basic Gas Laws

The state of a gas is characterized by its temperature, pressure and volume. If the temperature of a gas is 0°C* and the pressure 760 mm Hg, the conditions are referred to as standard.** The volume occupied by a gas in these conditions is designated by V_0, and the pressure by p_0.

All gases obey the following laws:

1. *At constant temperature, the volume of a mass of gas is inversely proportional to the pressure* (**Boyle-Mariotte law**):

$$\frac{p}{p_1} = \frac{V_1}{V} \quad \text{or} \quad pV = \text{const}$$

2. *At constant pressure, the volume of a mass of gas increases by 1/273 of the volume it occupies at 0°C for each degree rise in temperature* (**Gay-Lussac's law**):

$$V = V_0 + \frac{V_0 t}{273} = V_0 \left(\frac{273 + t}{273} \right)$$

where t is the temperature in Celsius degrees.

By introducing into this formula absolute temperature T, read from the absolute zero, i.e. the point marked by 273°C*** below zero on the centigrade scale $(T = 273 + t)$, we have another expression of the same law:

$$V = \frac{V_0 T}{273} \quad \text{or} \quad \frac{V}{T} = \frac{V_0}{273}$$

---

* Centigrade degrees should preferably be referred to as "Celsius degrees". In both cases the abbreviation is "°C".

** Standard temperature (0° C) and standard pressure (760 mm Hg) will be designated further in this textbook as STP.

*** To be more exact. 273.15°C.

Since $\frac{V_0}{273}$ is the constant for the given mass of a gas, the following equations hold true:

$$\frac{V}{T} = \text{const} \quad \text{or} \quad \frac{V}{T} = \frac{V_1}{T_1}$$

At constant pressure, the volume of gas changes in direct proportion to the absolute temperature.

If the volume of gas remains constant, for example during heating in a sealed vessel, *the gas pressure changes in direct proportion to the absolute temperature:*

$$\frac{p}{T} = \frac{p_1}{T_1}$$

By employing these laws, one can calculate:

(1) the changes in gas pressure with varying volume and temperature of a gas; and

(2) the changes in gas volume with varying pressure and temperature*.

Example 1. At a given temperature, the pressure of a gas occupying a volume of 3 litres is 700 mm Hg. What will be the pressure if the volume is reduced to 2.8 litres without changing the temperature?

Solution. The product of a gas volume and its pressure is a constant value at a given temperature. If we designate the sought pressure as p, the following can be written:

$$p \times 2.8 = 700 \times 3$$

whence

$$p = \frac{700 \times 3}{2.8} = 750 \text{ mm Hg}$$

By expressing the obtained result in units of pressure accepted in the International System (newtons per sq m) and knowing that a pressure of 1 mm Hg is 133.32 N/sq m (see Appendix) we obtain

$$p = 750 \times 133.32 = 99{,}990 \text{ N/sq m}$$

---

* Only the so-called ideal gases obey these basic gas laws. In the imaginary ideal gas, the molecules are only points which have no volume and do not act on each other. The behaviour of an ideal gas somewhat differs from that of a real gas, the distinction being more apparent at lower temperatures and greater pressures. Therefore at great pressures or low temperatures the calculations with the gas laws are only tentative.

Example 2. At a temperature of 27°C, a gas has a volume of 600 ml. What will be the volume of the gas at 57°C, if the pressure remains unaltered?

Solution. Let us designate the sought volume by V_1 and the corresponding temperature by T_1. Then

$$\frac{V}{T} = \frac{V_1}{T_1}$$

According to the condition of the problem $V = 600$ ml, $T = 273 + 27 = 300°K*$ and $T_1 = 273 + 57 = 330°K$. By substituting these values into the formula we have

$$\frac{600}{300} = \frac{V_1}{330}$$

whence

$$V_1 = \frac{600 \times 330}{300} = 660 \text{ ml}$$

Example 3. At 15°C oxygen pressure inside a cylinder is 90 atm. At what temperature will it be 100 atm?

Solution. The problem can be solved as in the previous case. Let the sought temperature be T_1. Then $T = 273 + 15 = 288°K$, $p = 90$ atm, $p_1 = 100$ atm. By substituting these values into equation

$$\frac{p}{T} = \frac{p_1}{T_1} \quad \text{or} \quad T_1 = \frac{p_1 \cdot T}{p}$$

we have

$$T_1 = \frac{100 \times 288}{90} = 320° \text{ K or } 47°\text{C}$$

Example 4. A gas occupies a volume of 152 ml at 25°C and a pressure of 745 mm Hg. What volume will it occupy at 0°C and 760 mm Hg (at STP)?

Solution. First let us calculate the volume V_1 which the gas will occupy at a pressure raised to 760 mm Hg if the temperature remains unchanged:

$$V_1 \times 760 = 152 \times 745$$

$$V_1 = \frac{152 \times 745}{760} = 149 \text{ ml}$$

---

* °K designates temperature degrees as read off from absolute zero.

Thus, at standard pressure and a temperature of 25°C the gas will occupy a volume of 149 ml.

Now reduce the temperature to 0°C leaving the pressure unchanged, and find the new volume of the gas (V_0).

Since $V_1 = 149$ ml, $T_1 = 273 + 25 = 298°K$ and $T_0 = 273°K$,

$$\frac{V_0}{273} = \frac{149}{298}$$

whence

$$V_0 = \frac{273 \times 149}{298} = 136.5 \text{ ml}$$

* * *

The relationship between the volume occupied by a gas, the pressure and the temperature would be usually expressed by a general equation which combines the Boyle-Mariotte and Gay-Lussac laws:

$$pV = \frac{p_0 V_0 T}{273}$$

where p is the pressure and V the volume of a gas at a given temperature T,

p_0 is the pressure and V_0 the volume of a gas at STP.*

This equation can be used for determining any value provided the others are known. For example, V_0 can be determined from the formula:

$$V_0 = \frac{pV \times 273}{p_0 T}$$

By employing this formula one can calculate any volume of a gas at standard conditions provided the temperature and pressure are known. For example, by substituting the data from the previous problem one has

$$V_0 = \frac{745 \times 152 \times 273}{760 \times 298} = 136.5 \text{ ml}$$

PROBLEMS

27. The pressure of a gas occupying a volume of 2.5 litres is 1.2 atm. What will be the pressure if the volume is reduced to 1 litre, the temperature remaining constant?

_____

* This equation can be easily derived from the previous problem when solving it in the general form and using letter designations.

28. A gas occupies a volume of 30 litres at a pressure of 2 atm. What pressure is required to reduce the initial volume to 25 litres, the temperature remaining constant?

29. One gram of air at STP has a volume of 773 ml. What volume will it occupy at 0°C and a pressure of 700 mm Hg?

30. A steel cylinder of 12-litre capacity contains oxygen at 0°C and a pressure of 150 atm. What will be the volume of the oxygen reduced to STP?

31. At 0°C and 896 mm Hg, 28 g of nitrogen occupy a volume of 19 litres. What volume will they occupy at STP?

32. A gas occupies a volume of 580 ml at 17°C. What volume will it occupy at 100°C, the pressure remaining constant?

33. A gas at 0°C is sealed in a vessel. To what temperature should the gas be heated to double the initial pressure?

34. A gas occupies a volume of 150 ml at 27°C. To what must the temperature be changed in order to increase its volume to 200 ml, the pressure remaining constant?

35. The pressure on a gas in a sealed vessel at 12°C is 750 mm Hg. What will the pressure be if the vessel is heated to 30°C?

36. At 7°C the gas pressure in a closed vessel is 720 mm Hg. To what value will the pressure be changed if the vessel is cooled to —33°C?

37. The pressure on a gas in a sealed vessel at 21°C is 840 mm Hg. To what temperature must the gas be cooled to reduce the pressure to standard?

38. The temperature of nitrogen compressed in a steel cylinder to 150 atm is 18°C. The maximum allowable pressure in the cylinder is 200 atm. At what temperature will nitrogen attain this pressure?

39. A certain sample of a gas under a pressure of 720 mm Hg and a temperature of 15°C occupies a volume of 912 ml. What volume will the gas occupy at standard conditions?

40. A gas occupies a volume of 608 ml at a pressure of 740 mm Hg and a temperature of 91°C. What volume will the gas occupy at standard conditions?

41. A gas occupies a volume of 5 litres at a temperature of 27°C and a pressure of 720 mm Hg. What volume will the same sample occupy at 39°C and 780 mm Hg?

42. On reaction of 1.28 g of metal with water, 380 ml of hydrogen measured at 21°C and 784 mm Hg evolved. Determine the equivalent weight of the metal.

2. Partial Pressure of Gas

The pressure of a gas on the walls of a vessel is due to the action of its molecules as they strike the walls. Therefore, at constant temperature, the pressure of a gas is directly proportional to the number of molecules contained in a unit volume.

If a vessel contains a mixture of gases, each gas exerts a pressure as if it alone occupied the entire container. The part of the total pressure exerted by a gaseous mixture which is due to the gas in question is called the *partial pressure* of a gas.

The total pressure of a gaseous mixture is equal to the sum of partial pressures of individual gases in the mixture (**Dalton's law**).

If partial pressures of individual gases in a mixture are known, the total pressure of a gaseous mixture can easily be calculated. And conversely, if volumes of individual gases in the mixture and the total pressure of the mixture are known, partial pressures of individual gases can be calculated.

Example 1. Two litres of oxygen and 4 litres of sulphurous anhydride are mixed. Both gases have the same pressure of 750 mm Hg. The volume of the mixture is 6 litres. Determine the partial pressure of each gas in the mixture.

Solution. On mixing with sulphurous anhydride, the volume of the oxygen has increased $6 : 2 = 3$ times, whereas the volume of the anhydride $6 : 4 = 1.5$ times. The values of pressures of the component gases reduced accordingly. Hence, the partial pressures are as follows:

oxygen $750 : 3 = 250$ mm Hg or 33,330 N/sq m

sulphurous anhydride $750 : 1.5 = 500$ mm Hg or 66,660 N/sq m

This example clearly shows that the partial pressures of gases are proportional to their volume content of the mixture.

Example 2. Three litres of carbon dioxide, 4 litres of oxygen and 6 litres of nitrogen are mixed. Pressure of carbon dioxide before mixing was 720 mm Hg, oxygen pressure was 810 and nitrogen 680 mm Hg. The volume of the mixture is 10 litres. Determine its pressure.

Solution. By analogy with the previous example, first determine partial pressures of individual gases:

Pressure of carbon dioxide: $\frac{720 \times 3}{10} = 216$ mm Hg or 28,800 N/sq m

Pressure of oxygen: $\frac{810 \times 4}{10} = 324$ mm Hg or 43,200 N/sq m

Pressure of nitrogen: $\frac{680 \times 6}{10} = 408$ mm Hg or 54,400 N/sq m

The total pressure of the gas mixture (p) is equal to the sum of partial pressures of component gases:

$$p = 216 + 324 + 408 = 948 \text{ mm Hg or } 126,400 \text{ N/sq m}$$

Example 3. A mixture of gases contains 30 per cent of methane and 70 per cent (v/v) of hydrogen. The pressure of the mixture is 800 mm Hg. Determine the partial pressure of each gas separately.

Solution. Since partial pressure of a gas is proportional to the gas content (v/v) of a mixture, 800 mm should be divided proportionally to 30 and 70:

Methane pressure $\frac{800 \times 30}{100} = 240$ mm Hg or 32,000 N/sq m

Oxygen pressure $\frac{800 \times 70}{100} = 560$ mm Hg or 74,660 N/sq m

PROBLEMS

43. A gaseous mixture is prepared of 2 litres of hydrogen at a pressure of 700 mm Hg and 5 litres of methane at 840 mm Hg. The total volume of the mixture is the sum of the individual volumes. What are the partial pressures of the gases in the mixture and the total pressure?

44. Air contains 21 per cent of oxygen (by volume). What is the partial pressure of oxygen in the air at standard pressure?

45. 7.4 litres of oxygen are placed above water in a gasometer at 23°C and 781 mm Hg. The pressure of water vapour at

23°C is 21 mm Hg. What volume will the oxygen occupy in the gasometer at STP?*

46. Three litres of nitrogen at a pressure of 720 mm Hg are mixed with 2 litres of oxygen. The volume of the mixture is 5 litres, and the total pressure is 732 mm Hg. What was the original pressure of the oxygen?

47. How many litres of nitrogen at 756 mm Hg should be added to 2 litres of oxygen at the same pressure to adjust the partial pressure of the oxygen in the mixture to 630 mm Hg?

48. A gaseous mixture consists of nitrogen monoxide and dioxide. Calculate the volume per cent content of the gases in the mixture if their partial pressures are 272 and 528 mm Hg, respectively.

49. Two oxygen cylinders of 3- and 4-litre capacity are interconnected through a pipe with a cock. With the stopcock closed, the pressure inside the first cylinder is 420 mm Hg, and the oxygen pressure inside the other cylinder is 777 mm Hg. The temperature of the gas is the same in both cylinders. What will the pressure in the cylinders be at the same temperature when the stopcock is opened?

50. A gaseous mixture is prepared from 3 litres of methane at a pressure of 720 mm Hg, 4 litres of hydrogen at a pressure of 630 mm Hg and 1 litre of carbon monoxide at a pressure of 816 mm Hg. The mixture occupies a volume of 8 litres. Determine the partial pressures of individual gases and the total pressure of the mixture.

3. Avogadro's Law. Reacting Volumes of Gaseous Substances

Equal volumes of any gases at the same temperature and the same pressure contain equal numbers of molecules (**Avogadro's law**).

In other words, if equal volumes of various gases at the same temperature contain equal numbers of molecules, the pressure of these gases will also be equal. It follows therefore that at constant volume and constant temperature the pressure of

---

* In solving the problem, bear in mind that the pressure in the gasometer is the sum of partial pressures of oxygen and saturated water vapour.

a gas depends only on the number of molecules and does not depend on their nature.

The volumes of interacting gases are always in a ratio of small whole numbers to each other and to the volumes of the gaseous reaction products (the **law of combining volumes,** or **Gay-Lussac's "Chemical" law**).

When the volume ratio of the reacting gases is known one can establish the composition of molecules of the resultant product. And conversely, if the composition of the molecules of the starting substances and of the reaction products is known, one can easily calculate the volume ratio of the reacting gases, provided the temperature and the pressure remain unaltered or else the change in the pressure due to the reaction can be calculated (if the reaction is carried out in a sealed vessel at constant temperature).

Example 1. Two volumes of nitrogen oxide NO combine with one volume of oxygen to produce two volumes of a new gas. Determine the formula of the new gas.

Solution. Since equal volumes of gases (the other conditions being the same) contain equal numbers of molecules, it follows that each pair of nitrogen oxide molecules combines with one molecule of oxygen to produce two molecules of a new gas.

If we designate the composition of the molecules of the new gas by the formula N_xO_y the reaction will be expressed as this:

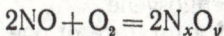

$$2NO + O_2 = 2N_xO_y$$

The calculation of nitrogen and oxygen atoms in the left part of the equation shows that two molecules of the new gas should contain two atoms of nitrogen and four atoms of oxygen, that is the molecule of the new gas consists of one atom of nitrogen and two atoms of oxygen. The formula of the new gas will therefore be NO_2.

Example 2. When burned, methane CH_4 forms carbon dioxide and water vapour. How do the volumes of the reacting gases compare if they are measured in the same conditions?

Solution. When burned completely, methane reacts with two molecules of oxygen to produce one molecule of carbon dioxide and two molecules of water vapour:

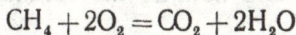

$$CH_4 + 2O_2 = CO_2 + 2H_2O$$

This means that one volume of methane and two volumes of oxygen produce two volumes of water vapour and one volume of carbon dioxide, measured in the same conditions.

Example 3. A mixture of equivalent quantities of hydrogen and oxygen is present in a sealed vessel at a temperature above 100°C. How will the pressure inside the vessel change if the mixture is exploded and the temperature of the vessel contents is then reduced to the initial value?

Solution. Two hydrogen atoms and one oxygen atom are required to produce one molecule of water. It follows therefore that during the reaction between oxygen and hydrogen, each pair of hydrogen molecules and one oxygen molecule produce two molecules of water or water vapour. Thus, the total number of molecules reduces in the reaction 1.5 times. Since the reaction is effected in a closed vessel, in other words, the volume remains constant, and since in the end of the reaction the temperature of the vessel contents is restored to the original level, the 1.5-fold decrease of the number of molecules will reduce the pressure inside the vessel accordingly.

PROBLEMS

51. One volume of carbon monoxide and one volume of chlorine react to produce one volume of phosgene. Determine the formula of phosgene.

52. On explosion of a mixture of one volume of a gas and one volume of hydrogen, one volume of water vapour and one volume of nitrogen were formed. All measurements were made in the same conditions. Determine the formula of the gas.

53. When burned, one volume of a gaseous hydrocarbon formed three volumes of carbon dioxide and four volumes of water vapour. All volumes were measured in the same conditions. Determine the formula of the hydrocarbon.

54. Phosphorus vapour reacts with oxygen in the ratio of 1 : 5 by volume to produce phosphoric anhydride. How many atoms does a molecule of phosphorus vapour contain?

55. When ammonia is burned in chlorine, hydrogen chloride and nitrogen are produced. What volumes of ammonia and chlorine react and how do the volumes of the resultant gases compare?

56. On oxidation of ammonia in the presence of a catalyst, nitrogen oxide NO and water vapour are produced. In what volume ratio do ammonia and oxygen interact in this reaction and how do the volumes of the obtained gaseous products compare?

57. In what volume ratio do acetylene C_2H_2 and oxygen react when acetylene is burned?

58. How many litres of carbon dioxide are produced on burning two litres of butane C_4H_{10}? The volumes of both gases are measured in the same conditions.

59. On burning, ethylene C_2H_4 produced 15 litres of carbon dioxide. How many litres of ethylene were burned?

60. How many litres of oxygen are required to burn 10 litres of hydrogen sulphide H_2S, if the reaction products are sulphurous anhydride and water?

61. After explosion of 20 ml of mixture of hydrogen with oxygen, 3.2 ml of oxygen remained unreacted. Determine and express in per cent (v/v) the original composition of the mixture.

62. One volume of nitrogen oxide NO is mixed with two volumes of oxygen. All nitrogen oxide is converted into nitrogen dioxide NO_2. Determine volume percentage of each component in the mixture.

63. On passing a mixture of equal volumes of sulphurous anhydride and oxygen through a contact apparatus 90 per cent of sulphurous anhydride SO_2 are converted into sulphuric anhydride SO_3. Determine the volume ratio of the gases in the mixture released from the contact apparatus.

64. A mixture of equal volumes of hydrogen and oxygen is present in a steel cylinder at a pressure of 0.2 atm. The temperature of the mixture is 150°C. What will the gas pressure be after explosion of the mixture and after reduction of the temperature to the original level?

65. Calculate the change in the pressure of a gaseous mixture in a sealed vessel consisting of one volume of nitrogen and three volumes of hydrogen, if half the total nitrogen converts into ammonia.

66. A mixture of three volumes of oxygen and one volume of methane is present in a sealed vessel at a temperature of 120°C and a pressure of 0.5 atm. Calculate the pressure inside

the vessel after explosion of the mixture and after reduction
of the temperature to the original level.

67. A mixture of equivalent quantities of oxygen and acetylene C_2H_2 is present in a closed vessel at 150°C. Calculate
the change in the pressure inside the vessel if the mixture is
exploded and the temperature of the vessel content is reduced
to the original level.

68. A mixture contained in a closed vessel and consisting
of three volumes of chlorine and one volume of hydrogen was
exposed to scattered light at constant temperature. After
a certain exposure, the chlorine content of the mixture reduced 20 per cent. Has the pressure inside the vessel changed?
What is the new ratio of the gas volumes in the mixture?
Express it in per cent.

4. Gram-Molecular Volume of Gas

In order to measure the quantity of a substance, besides
the gram and kilogram units of mass, a specific chemical measure of quantity of a substance is employed in chemistry called
the gram-molecule or mole, for short.

*A gram-molecule is a quantity of a substance, the mass of
which in grams is numerically equal to the molecular weight
of the substance.*

The convenience of this measure is that the gram-molecules
of all substances, though having different masses, contain
equal number of molecules, namely **6.02 × 10²³ (Avogadro's
number).** It follows therefore that the gram-molecules of all
gaseous substances must occupy equal volumes at equal
conditions.

A volume occupied by one gram-molecule of a gas (gram-
molecular volume of a gas) at STP is **22.4 litres.**

If the gram-molecular volume of a gas and its molecular
weight are known, one can easily calculate the volume and
the pressure of any quantity of the gas in any conditions.

Example 1. Determine the volume occupied by 5.25 g of
nitrogen at 26°C and a pressure of 736 mm Hg.

Solution. From the gram-molecular volume of nitrogen and
the molecular weight of the gas which is 28 we can determine
the volume at STP (V_0) which the gas will occupy:

$$5.25:28 = V_0:22.4$$
$$V_0 = \frac{5.25 \times 22.4}{28} = 4.2 \text{ litres}$$

Now recalculate the obtained volume for the conditions specified in the problem:

$$V = \frac{p_0 V_0 T}{p\,273} = \frac{760 \times 4.2 \times 299}{736 \times 273} = 4.75 \text{ litres}$$

Example 2. A closed vessel of 500 ml capacity contains 0.75 g of nitrogen oxide NO. The temperature of the gas is 39°C. Determine its pressure.

Solution. The gram-molecule of any gas occupying at 0°C the volume of 22.4 litres produces a pressure of 1 atm. Taking this into account and bearing in mind that the mass of a gram-molecule of NO is 30 g, first determine the pressure produced by 0.75 g of nitrogen oxide occupying the same volume. Since at constant volume and temperature the gas pressure is directly proportional to its quantity (the number of molecules), the following proportion can be used to determine the sought pressure:

$$p:1 = 0.75:30$$

whence

$$p = 0.025 \text{ atm}$$

Now applying the basic gas laws one can determine:
(1) pressure (p_1) produced by the 0.75 g of nitrogen oxide at 0°C in the volume of 0.5 litre:

$$p_1 \times 0.5 = 0.025 \times 22.4$$
$$p_1 = \frac{0.025 \times 22.4}{0.5} = 1.12 \text{ atm}$$

(2) pressure (p_2) produced by the gas at 39°C:

$$\frac{p_2}{273 + 39} = \frac{1.12}{273}$$
$$p_2 = \frac{312 \times 1.12}{273} = 1.28 \text{ atm}$$

Hence the sought pressure is 1.28 atm, or 129,700 N/sq m*.

---

* 1 atm = 101,325 N/sq m.

Example 3. What volume will one gram-equivalent of oxygen occupy at 0°C and 700 mm Hg?

Solution. A gram-equivalent of oxygen weighs 8 g. Since a gram-molecule of oxygen, that is 32 g of oxygen, at STP occupies a volume of 22.4 litres, the gram-equivalent of oxygen at STP will occupy the volume of

$$\frac{22.4 \times 8}{32} = 5.6 \text{ litres}$$

Now determine the volume which the oxygen will occupy at 700 mm Hg:

$$V \times 700 = 5.6 \times 760$$

$$V = \frac{5.6 \times 760}{700} = 6.08 \text{ litres}$$

PROBLEMS

69. Calculate the volumes that 80 g of oxygen, 3 g of nitrogen oxide and 128 g of sulphurous anhydride occupy at STP.

70. It is known that 1 g of water at 100°C occupies a volume of approximately 1 ml. What volume will the same water occupy when converted into vapour at the same temperature?

71. Calculate the volumes occupied at STP by gram-molecules of oxygen, carbon dioxide, glycerol $C_3H_8O_3$ (density 1.26 g/cu cm) and water.

72. Determine the volume occupied by 70 g of nitrogen at a temperature of 21°C and a pressure of 1.4 atm.

73. The answer to the question, what volume the gram-molecule of water occupies at STP, was 22.4 litres. Why is this answer incorrect and what is the correct answer?

74. What volume will 1 g of hydrogen, oxygen and carbon dioxide occupy at STP?

75. Potassium chlorate $KClO_3$ decomposes on heating into potassium chloride KCl and oxygen. How many litres of oxygen at 0°C and 760 mm Hg can be obtained from one mole of potassium chlorate?

76. Equal weights of oxygen, hydrogen and methane are taken at equal conditions. How do the volumes of these gases compare?

77. What volumes do 11 kg of carbon dioxide, 1 kg of hydrogen and 16 kg of methane occupy at STP?

78. Determine the volume occupied by one gram-molecule of oxygen at 20°C and standard pressure.

79. How many gram-molecules does 1 cu m of any gas contain at STP?

80. What is the volume that 27×10^{21} molecules of gas occupy at STP?

81. How many molecules does 1 ml of hydrogen contain at STP?

82. What volume do 4 g of methane occupy at STP? What quantity of hydrogen will occupy the same volume at STP?

83. Is the number of molecules in 1 g of hydrogen equal to that in 1 g of oxygen? in one litre of hydrogen and one litre of oxygen in the same conditions? in the gram-molecule of hydrogen and gram-molecule of oxygen?

84. There are one litre of oxygen at 0°C and 760 mm Hg and one litre of carbon dioxide at 27°C and 900 mm Hg. Which gas contains a greater number of molecules? How do these quantities compare?

85. How many molecules does 1 ml of water contain at 4°C?

86. How many atoms of mercury does 1 ml of Torricellian vacuum contain at 0°C if the mercury vapour pressure at this temperature is 0.0002 mm Hg? Mercury vapour is monoatomic.

87. A sealed vessel of 1-litre capacity contains 0.25 g of nitrogen at 0°C. What is the gas pressure?

88. A vessel of 1-litre capacity holds 0.05 mole of a gas at 0°C. Determine the gas pressure.

89. What will be the pressure of 6 g of nitrogen oxide NO in a sealed vessel of 2-litre capacity at 0°C?

90. What is the atmospheric pressure on the top of Kazbek summit (Caucasus range), if the mass of 1 litre of air taken there at 0°C is 0.7 g? At STP, 1 litre of air weighs 1.29 g.

91. A steel cylinder of 10-litre capacity holds 1 kg of oxygen at 0°C. What is the pressure of the oxygen inside the cylinder?

92. A vessel of 7-litre capacity contains 0.4 g of hydrogen and 3.15 g of nitrogen at 0°C. Determine the partial pressure of hydrogen and the total pressure of the mixture.

5. Mass of Gas. Density of Gas.

The mass of any gas taken in any conditions can be calculated either from the mass of 1 litre of the gas at STP* or from the molecular weight of the gas.

Example 1. The mass of 1 litre of oxygen at STP is 1.43 g. Determine the mass of 608 ml of oxygen at 25°C and a pressure of 745 mm Hg.

Solution. First the given volume should be calculated as at STP:

$$V_0 = \frac{pV \times 273}{p_0 T} = \frac{745 \times 608 \times 273}{760 \, (273 + 25)} = 546 \text{ ml or } 0.546 \text{ litre}$$

Now the mass (m) of the calculated volume of oxygen can be determined:

$$m = 1.43 \times 0.546 = 0.78 \text{ g}$$

Example 2. What is the mass of 1 litre of carbon monoxide at STP?

Solution. The molecular weight of carbon monoxide is 28. It follows that at STP the mass of 22.4 litres of carbon monoxide is 28 g. Hence 1 litre of carbon monoxide is:

$$28 : 22.4 = 1.25 \text{ g}$$

Example 3. What is the mass of 4 litres of carbon dioxide at 0°C and a pressure of 665 mm Hg?

Solution. Reduce the volume of carbon dioxide to STP conditions:

$$V_0 \times 760 = 4 \times 665$$
$$V_0 = \frac{4 \times 665}{760} = 3.5 \text{ litres}$$

The mass of one gram-molecule of carbon dioxide is 44 g. Designate the sought mass as m and make out the proportion:

$$m : 44 = 3.5 : 22.4$$

whence

$$m = \frac{44 \times 3.5}{22.4} = 6.88 \text{ g}$$

---

* The required data are available in reference books.

The ratio of the mass of a given gas to the mass of the same volume of another gas at the same temperature and pressure is called the density of the first gas with respect to the second.

Let the density of a gas be d, the mass of the first gas m_1 and of the second gas m_2; then

$$d = \frac{m_1}{m_2}$$

As a rule, the density of a gas is determined with respect to hydrogen or air. For example, if the mass of one litre of sulphurous anhydride at 0°C and 760 mm Hg is 2.88 g, the mass of 1 litre of hydrogen is 0.09 g and that of 1 litre of air is 1.29 g, the density of sulphurous anhydride can be calculated as follows:

with respect to hydrogen

$$d_{H_2} = \frac{2.88}{0.09} = 32$$

with respect to air

$$d_{air} = \frac{2.88}{1.29} = 2.23$$

There is a very simple relationship between the density of a gas and its molecular weight: since equal volumes of two gases in the same conditions contain equal number of molecules, their masses should relate to each other as the molecular weights. Designate the mass of the first gas as m_1, of the second gas as m_2, the molecular weight of the first gas as M_1 and of the second gas as M_2. Then:

$$\frac{m_1}{m_2} = \frac{M_1}{M_2}$$

But $\frac{m_1}{m_2}$ is the density of the first gas with respect to the second, which we already designated as d. It follows therefore that

$$d = \frac{M_1}{M_2}$$

The derived formula can be used for calculating density of a given gas with respect to another, provided the molecular weights of both are known. If it is necessary to calculate the density of a gas with respect to air, which is a mixture of

several gases, the mean molecular weight of air should be
known, which can be calculated from its density with respect
to hydrogen. It is known to be 29.

Thus, the density of a gas with respect to air is

$$d_{air} = \frac{M_1}{29}$$

If d_{air} is greater than 1, the gas is heavier than air and vice
versa, if d_{air} is less than 1, the gas is lighter than air.

Example 4. Determine the density of ammonia with respect to air.

Solution. The molecular weight of ammonia NH_3 is 17;
the mean molecular weight of air is 29. Hence, the density
of ammonia with respect to air is

$$d_{air} = \frac{17}{29} = 0.59$$

Example 5. What is the density with respect to hydrogen
of a gaseous mixture containing 75 per cent (v/v) of methane
and 25 per cent (v/v) of oxygen?

Solution. From the formulas of methane (CH_4) and oxygen
(O_2) it follows that the molecular weight of methane is 16
and that of oxygen is 32.

In compliance with the conditions of the problem the gaseous mixture contains 25 molecules of oxygen per each 75 molecules of methane, therefore its mean molecular weight is

$$\frac{16 \times 75 + 32 \times 25}{100} = 20$$

Since the molecular weight of hydrogen is 2, the density
of the gaseous mixture with respect to hydrogen is

$$d_{H_2} = \frac{20}{2} = 10$$

PROBLEMS

93. On decomposition of calcium carbonate, 855 ml of
carbon dioxide at 750 mm Hg and 12°C were obtained. Determine the mass of the gas.

94. Knowing the mean molecular weight of air, what is the
mass of its 1 litre?

95. Determine the mass of 1 litre of air at a temperature of 51°C and a pressure of 700 mm Hg if at STP the mass of one litre of air is 1.29 g.

96. Determine the mass of 20 litres of chlorine at STP.

97. A steel cylinder of 20-litre capacity holds hydrogen. At 12°C its pressure is 125 atm. Determine the mass of the hydrogen.

98. A gasometer of 20-litre capacity is filled with hydrogen. The pressure of the gas is 747 mm Hg, and the temperature is 27°C. Calculate the mass of the hydrogen, bearing in mind that the pressure of water vapour at this temperature is 27 mm Hg.

99. A steel cylinder of 14-litre capacity holds oxygen. The gas temperature is 0°C and the pressure is 80 atm. What is the mass of the oxygen?

100. Determine the mass of 1 litre of benzene C_6H_6 vapour at standard pressure and a temperature of 117°C.

101. The density of a gas with regard to hydrogen is 17. What is the mass of 1 litre of this gas at STP? What is its density with respect to air?

102. The density of the lighting gas with respect to air is 0.4. What is the mass of 1 cubic metre of this gas at STP?

103. Determine the mass of air filling the room of $5 \times 4 \times 3$ m at standard pressure and a temperature of 18°C.

104. Calculate the density with respect to air and the mass at STP of one litre of nitrogen N_2, carbon monoxide CO, fluorine F_2 and hydrogen sulphide H_2S.

105. Determine the mass of 1 litre of hydrogen bromide HBr at STP. What is the density of this gas with respect to air?

106. Determine the mass of one litre of phosgene at STP, and its density with respect to air and hydrogen. The formula of phosgene is $COCl_2$.

107. Is water vapour lighter or heavier than air? How do their densities relate?

108. Determine the mass of 1 litre of carbon dioxide at STP from its formula. What is the gas density with respect to hydrogen and air?

109. Determine, which of the gases are lighter than air: fluorine F_2, ammonia NH_3, methane CH_4, nitrogen dioxide

NO_2, carbon monoxide CO. How do their densities relate to that of the air?

110. What is the density with respect to air of water gas which is a mixture of equal volumes of hydrogen and carbon monoxide?

111. The density of a gas with respect to air is 1.52. What volume will 5.5 g of this gas occupy at STP?

112. Determine the density with respect to hydrogen and air of methane CH_4, nitrous oxide N_2O, ethyl alcohol vapour C_2H_5OH, diethyl ether vapour $(C_2H_5)_2O$.

113. Determine the density with respect to air of lighting gas of the following composition (per cent by volume): 48% hydrogen H_2, 35% methane CH_4, 8% carbon monoxide CO, 4% ethylene C_2H_4, 2% carbon dioxide CO_2 and 3% nitrogen N_2.

6. Molecular Weight of a Substance in the Gaseous State

To calculate the molecular weight of a substance in the gaseous state, it is necessary to know the mass, volume, pressure and temperature of a certain quantity of this substance. If these are known, one may calculate the density of the gaseous substance and hence its molecular weight, or find the molecular weight directly from the gram-molecular volume.

A. Calculation of Molecular Weight of a Gas or Vapour from Its Density

It follows from the formula expressing the relationship between the density of a gas and its molecular weight (see Sec. 5) that

$$M_1 = dM_2$$

i.e. *the molecular weight of a gas is equal to its density with respect to another gas multiplied by the molecular weight of the latter.*

Example. The density of a gas with respect to air is 1.17. Determine its molecular weight.

Solution. By substituting into the above equation 1.17 for d and the mean molecular weight of air 29 for M_2 we obtain

$$M_1 = 1.17 \times 29 = 34$$

B. Determination of Molecular Weight of a Gas or Vapour from Its Gram-Molecular Volume

Since at standard conditions the gram-molecule of any gas (or vapour) occupies a volume of 22.4 litres, the sought molecular weight of the given gas, equal numerically to its gram-molecule, can be determined by calculating the mass of the given gas occupying the volume of 22.4 litres.

Example. At 27°C and a pressure of 800 mm Hg the mass of 380 ml of a gas is 0.455 g. Determine its molecular weight.

Solution. Reduce the given volume to standard conditions:

$$V_0 = \frac{800 \times 380 \times 273}{760 \times 300} = 364 \text{ ml or } 0.364 \text{ litre}$$

It follows therefore that the mass of 0.364 litre of the gas equals 0.445 g. The mass of the gas occupying 22.4 litres is

$$0.364 : 22.4 = 0.455 : x$$

$$x = \frac{22.4 \times 0.455}{0.364} = 28 \text{ g}$$

The molecular weight of the gas is 28.

PROBLEMS

114. The density of ethylene with respect to oxygen is 0.875. Determine the molecular weight of ethylene.

115. At a certain temperature, the density of sulphur vapour with respect to nitrogen is 9.14. How many atoms does a molecule of sulphur contain at this temperature?

116. How many atoms does a molecule of mercury vapour contain if the vapour density of mercury with respect to air is 6.92? (Atomic weight of mercury is 200.6.)

117. The density of phosphorus vapour with respect to air is 4.28. How many atoms does a molecule of phosphorus vapour contain?

2*

118. The mass of 1 litre of ozone at STP is 2 143 g. Determine the molecular weight of ozone and its density with respect to air.

119. On weighing acetylene it was found that the mass of 200 ml at STP was 0.232 g. Determine the molecular weight of acetylene.

120. At STP, the mass of 250 ml of a gas is 0.903 g. Determine its density with respect to air and the molecular weight of the gas.

121. The mass of 1 litre of a gas at STP is 1.52 g, and that of nitrogen is 1.25 g. Calculate the molecular weight of the gas (a) from its density with respect to nitrogen, and (b) from the gram-molecular volume.

122. At 17°C and 780 mm Hg, the mass of 624 ml of a gas is 1.56 g. Calculate the molecular weight of the gas.

123. Determine the molecular weight of a substance knowing that the mass of 380 ml of its vapour at 97°C and a pressure of 740 mm Hg is 1.9 g.

7. Equation of State of Gas and Its Use in Calculations

In the equation $pV = \frac{p_0 V_0 T}{273}$, which was given earlier, the magnitude $\frac{p_0 V_0}{273}$ depends on both the mass of the given gas and its properties (the value of V_0 is different for equal masses of various gases). However, if this equation is related to a quantity of a gas equal to one gram-molecule, the magnitude $\frac{p_0 V_0}{273}$ will be equal for all gases, since the gram-molecule of any gas at STP occupies the same volume of 22.4 litres. In these conditions the value $\frac{p_0 V_0}{273}$ is called the *universal gas constant* and is denoted by the letter R. Introducing R into the above equation, we have the equation referring to one gram-molecule of gas:

$$pV = RT$$

This equation is known as the *equation of state of gas*.

For n gram-molecules of gas the equation is rearranged into the following expression:

$$pV = nRT$$

But the number of gram-molecules of gas is equal to its mass (m), expressed in grams, divided by the molecular weight of the gas (M):

$$n = \frac{m}{M}$$

By substituting $\frac{m}{M}$ for n in the previous equation, we have

$$pV = \frac{m}{M} RT$$

The derived equation can be used to calculate any of its magnitudes provided the others are known.

The numerical value of R depends on the units used to express the volume and the pressure.

If the volume V_0 is expressed in litres and the pressure p_0 in atmospheres, then

$$R = \frac{1 \times 22.4}{273} = 0.082 \, \frac{\text{lit} \cdot \text{atm}}{\text{deg} \cdot \text{mole}}$$

Expressing the volume in millilitres and the pressure in millimetres of mercury, we obtain 62,400 for the value of R:

$$R = \frac{760 \times 22,400}{273} \approx 62,400 \, \frac{\text{mm Hg} \cdot \text{ml}}{\text{deg} \cdot \text{mole}}$$

By expressing the volume and the pressure in SI system units, that is the volume in cubic metres, (cu m/mole) and the pressure in newtons per square metre (N/sq m), we have the following:

$$R = \frac{101,325 \times 0.0224}{273} = 8.314 \, \frac{\text{N} \cdot \text{m}}{\text{deg} \cdot \text{mole}}$$

or 8.314 J/(deg·mole) *.

It should be noted that the equation of state of a gas is recommended for use in cases where pressure and temperature of a gas are other than standard. Whenever the gas is at STP, its mass, volume or molecular weight can be easily calculated directly from the gram-molecular volume.

---

* Joule (J) is the unit of work (energy, heat) in the SI system (see Appendix 1).

Example 1. Calculate the molecular weight of benzene knowing that the mass of 600 ml of its vapour at a temperature of 87°C and a pressure of 624 mm Hg is 1.3 g.

Solution. From the equation of state of a gas referred to any quantity of gas, we determine M:

$$M = \frac{mRT}{pV}$$

By substituting into this formula the data specified in the problem we have the following:

$$M = \frac{1.3 \times 62,400 \times 360}{624 \times 600} = 78$$

Example 2. A steel cylinder of 20.5-litre capacity is filled with oxygen. At 17°C the pressure inside the cylinder is 87 atm. What is the mass of the oxygen?

Solution. According to the equation given above, the mass of the gas is

$$m = \frac{pVM}{RT}$$

In compliance with the conditions of the problem the pressure $p=87$ atm, the volume $V=20.5$ litres and the temperature $T=273+17=290$°C. The molecular weight of oxygen is 32. By substituting these data into the formula, and bearing in mind that the volume is expressed in litres and the pressure in atmospheres (hence $R=0.082$), we have the following solution:

$$m = \frac{87 \times 20.5 \times 32}{0.082 \times 290} = 2,400 \text{ g}$$

PROBLEMS

124. The mass of 640 ml of a gas at 39°C and 741 mm Hg is 1.73 g. What is its molecular weight?

125. Calculate the mass of 1 cubic metre of air at 17°C and 624 mm Hg.

126. Determine the mass of 1 cubic metre of carbon dioxide at 27°C and 1.5 atm.

127. A gasometer of 20-litre capacity is filled with lighting gas, whose density with respect to air is 0.4, the pressure,

1.025 atm, and the temperature, 17°C. Calculate the mass of the lighting gas.

128. What is the volume of 1 kg of air at a temperature of 17°C and a pressure of 1 atm?

129. Determine the volume occupied by 80 g of oxygen at a temperature of 17°C and a pressure of 1.5 atm.

130. The mass of a 750-ml flask filled with oxygen at 27°C is 83.3 g. The mass of the empty flask is 82.1 g. What is the pressure of oxygen inside the flask?

131. What volume will 1 kg of carbon dioxide CO_2 occupy at a temperature of —9°C and a pressure of 1.64 atm?

132. Determine the molecular weight of chloroform $CHCl_3$, knowing that the mass of 350 ml of its vapour at a temperature of 91°C and a pressure of 728 mm Hg is 1.34 g.

133. What is the molecular weight of acetone if the mass of 500 ml of its vapour at a temperature of 87°C and a pressure of 720 mm Hg is 0.93 g?

CHAPTER III

DETERMINATION OF ATOMIC WEIGHT
(ATOMIC MASS)

In determining the atomic weight (atomic mass) of an element 1/12 of the mass of the carbon atom is employed as the unit. This unit of measuring atomic weights is called the **carbon unit.**

One gram equals 6.02×10^{23} carbon units.

The atomic weight (atomic mass) of an element is the mass of its atom expressed in carbon units.

Hence *the molecular weight (molecular mass) of a substance is defined as the mass of its molecule expressed in carbon units.*

1. Determination of Atomic Weight of an Element From Molecular Weights of Its Compounds

To determine atomic weight by this method, it is necessary first to determine the molecular weights of as many compounds of this element as possible. As soon as the gravimetric composition of these compounds is known, one can calculate how many carbon units make up the quantity of the element contained in one molecule of each compound. The least of these quantities (all others being multiple of it) is assumed to be the atomic weight of the element under examination.

Example. Determine the atomic weight of phosphorus, if the composition and the molecular weights of the compounds listed below are as follows:

Compound	Molecu- lar weight	Phosphorus content, per cent
Phosphoric anhydride	284	43.66
Phosphine	34	91.18
Pyrophosphoric acid	178	34.83
Phosphorous anhydride	220	56.36

Solution. Calculate, how many carbon units equal the mass of phosphorus contained in one molecule of each of the named compounds.

The mass of a molecule of phosphoric anhydride is 284 carbon units, of which 43.66 per cent is the share of phosphorus. It follows therefore that the mass of phosphorus contained in the molecule of phosphoric anhydride is

$$\frac{284 \times 43.66}{100} = 124 \text{ carbon units}$$

Following this model, one can determine the quantity of phosphorus contained in molecules of the other compounds (in carbon units):

Phosphine	31
Pyrophosphoric acid	62
Phosphorous anhydride	124

From the obtained data it appears that the least quantity of phosphorus is contained in the molecule of phosphine, viz., 31 carbon units. This number is the atomic weight of phosphorus.

2. Determination of Atomic Weight from Atomic Heat

By atomic heat of an element is understood the product of its atomic weight and the specific heat* of the corresponding simple substance.

The atomic heat of the majority of simple solid substances is approximately the same and averages 6.3 cal/g-atom deg (the **Rule of Dulong and Petit**).

It follows therefore that the atomic weight of an element can be determined approximately by dividing 6.3 by the specific heat of the corresponding simple substance:

$$\text{atomic weight} = \frac{6.3}{\text{specific heat}}$$

---

* The specific heat is the amount of heat needed to raise the temperature of one gram of a substance one Celsius degree.

Example 1. Specific heat of zinc is 0.093 cal/g·deg. Calculate the approximate atomic weight of zinc (A).
Solution.

$$A = \frac{6.3}{0.093} = 67.7$$

The exact atomic weight of zinc is 65.38.

* * *

The described methods can only be used to determine approximate atomic weights of elements. For exact determinations one should know the equivalent weight of the element which can be found experimentally to a great accuracy. There is a certain relationship between the equivalent weight of an element and its atomic weight, namely: the atomic weight of an element equals its equivalent weight multiplied by the valence of the same element:

atomic weight = equivalent weight × valence

The valence is always a whole number. Hence, dividing the approximate atomic weight by the equivalent weight and rounding the result, one can find the valence of the element. Further, multiplying the equivalent weight by the valence, one obtains the exact magnitude of the atomic weight.

Example 2. Titanium oxide contains 59.95 per cent of titanium. Specific heat of titanium is 0.13 cal/g·deg. Determine the exact atomic weight of this element.

Solution. 1. Basing on the atomic heat, one can calculate the approximate atomic weight of titanium (A):

$$A = \frac{6.3}{0.13} = 48.5$$

2. Now the equivalent weight of titanium (E) can be found from the proportion:

$$59.95 : 40.05 = E : 8$$

$$E = \frac{59.95 \times 8}{40.05} = 11.975$$

3. The valence of titanium can be determined by dividing the approximate atomic weight by the equivalent weight:

$$\frac{48.5}{11.975} = 4.05$$

The rounded result, viz., 4, is the valence of titanium.

4. The exact atomic weight of titanium can now be determined by multiplying the equivalent weight by the valence:

$$11.975 \times 4 = 47.9$$

PROBLEMS

134. Express in carbon units the mass of carbon contained in molecules of the compounds listed below:

Compound	Molecular weight	Carbon content, per cent (w/w)
Glucose	180	40
Tartaric acid	150	32
Propyl alcohol	60	80

135. Aniline contains 77.4 per cent of carbon. The molecular weight of aniline is 93. Calculate how many carbon units the mass of carbon atoms in the aniline molecule weighs and determine the number of these atoms in a molecule.

136. The density of toluene vapour with reference to hydrogen is 46, and the weight percentage of carbon in toluene is 91.3. Determine the number of carbon atoms in a toluene molecule.

137. Determine the number of chlorine atoms in a molecule of chloroform if the density of chloroform vapour with respect to air is 4.12, chlorine content in chloroform is 89.1 per cent and the atomic weight of chlorine is 35.5.

138. Determine the atomic weight of vanadium from the composition and molecular weight of the compounds listed below:

Compound	Molecular weight	Vanadium content, per cent (w/w)
Vanadium oxide	67	76.1
Vanadic anhydride	182	56.0
Vanadic fluoride	108	47.2
Vanadic sulphide	198	51.5

139. Calculate the atomic weight of silicon from the composition and density with respect to hydrogen of the gaseous compounds listed below:

Compound	Density with respect to hydrogen	Silicon content, per cent (w/w)
Silane	16	87.5
Disilane	31	90.3
Trisilane	46	91.3
Silicon fluoride	52	26.9

140. Determine the atomic weight of sulphur from the composition and the density with respect to hydrogen of sulphur compounds listed below:

Compound	Density with respect to hydrogen	Sulphur content, per cent (w/w)
Hydrogen sulphide	17	94.1
Sulphurous anhydride	32	50.0
Carbon sulphide	38	84.2
Sulphur chloride	67.5	47.4
Carbon oxysulphide	30	53.3

141. By using the table of atomic weights, determine atomic heats of the following metals:

Metal	Al	Ca	Ni	Hg	Cd
Specific heat, cal/g·deg	0.22	0.16	0.11	0.033	0.06

142. Whose specific heat is greater, of lead or of tin? Prove your answer by calculations.

143. Determine the exact atomic weight of tungsten if its equivalent weight is 30.65 and the specific heat is 0.035 cal/g·deg.

144. The equivalent weight of a metal is 23.24; the specific heat is 0.09 cal/g·deg. Determine the exact atomic weight of the metal.

145. Nickel oxide contains 70.97 per cent of nickel. The specific heat of nickel is 0.11 cal/g·deg. Determine the exact atomic weight of nickel.

146. On oxidation of 2.28 g of metal, 3.78 g of oxide were formed. The specific heat of the metal is 0.25 cal/g·deg. Determine its exact atomic weight.

147. The chloride of a metal contains 44.76 per cent of chlorine. The specific heat of the metal is 0.074 cal/g·deg. Determine its exact atomic weight.

148. The sulphide of a metal contains 64.72 per cent of the metal, whose specific heat is 0.11 cal/g·deg. Determine its exact atomic weight, if the equivalent weight of sulphur is 16.

149. On dissolution of 2 g of a metal in sulphuric acid, 4.51 g of sulphate were formed. The specific heat of the metal is 0.057 cal/g·deg. What is the valence of the metal and its exact atomic weight?

150. 1.48 g of a metal displace from an acid 0.594 litre of hydrogen as measured at STP. The specific heat of the metal is 0.11 cal/g·deg. Determine its exact atomic weight.

151. On heating 2 g of a metal, 2.539 g of the oxide, in which the metal is tetravalent, were formed. What is the atomic weight of the metal?

152. The oxide of a pentavalent element contains 56.33 per cent of oxygen. What is the atomic weight of the element?

CHAPTER IV

DETERMINING THE FORMULA
OF A COMPOUND

The chemical formula of a compound shows what elements compose a given substance and how many atoms of each element are contained in its molecule.

The composition of a compound can be expressed by a simple, or empirical, formula, and by a true chemical, or molecular, formula.

The *empirical formula* expresses the simplest possible atomic composition of a molecule of a given substance corresponding to the weight ratios of the elements forming the given substance.

The *molecular formula* shows the actual number of atoms of each element in the molecule.

1. Derivation of Simplest Formulas

To derive an empirical formula of a compound, it is only sufficient to know its weight composition and atomic weights of the elements which form the compound.

Example 1. Derive the formula of chromium oxide knowing that it contains 68.4 per cent of chromium and 31.6 per cent of oxygen (atomic weight of chromium is 52).

Solution. Let the number of chromium atoms in the molecule of its oxide be x and that of oxygen atoms, y. Since the mass of the chromium atom is 52 carbon units, and that of the oxygen atom is 16 carbon units, the mass of all chromium atoms contained in the molecule is $52x$, and that of the oxygen atoms, $16y$. The ratio of these masses expresses the composition of the molecule of chromium oxide, and hence the composition of the substance. On the other hand, the same composition is expressed by the ratio 68.4 : 31.6.

Equating the ratios we get:

$$52x:16y = 68.4:31.6$$

Now we clear the unknown x and y of their coefficients by dividing the antecedents of the ratios by 52 and the consequents by 16:

$$x:y = \frac{68.4}{52}:\frac{31.6}{16} = 1.32:1.98$$

The obtained proportion shows that a molecule of chromium oxide contains 1.98 atoms of oxygen per 1.32 atoms of chromium. But the number of atoms in a molecule can only be an integer. Therefore, to express the ratio $x:y$ by integers, we divide both terms of the second ratio by the smaller of them:

$$x:y = \frac{1.32}{1.32}:\frac{1.98}{1.32} = 1:1.5$$

Now by multiplying both terms of the second ratio by 2 we obtain

$$x:y = 2:3$$

Thus, in a molecule of chromium oxide, two atoms of chromium combine with three atoms of oxygen. A series of formulas meet these conditions: Cr_2O_3, Cr_4O_6, Cr_6O_9, etc. Since the molecular weight of chromium oxide is unknown, it is impossible to determine which of these formulas expresses the true composition of the molecule. By selecting the smallest possible number of atoms for x and y ($x=2$ and $y=3$) we derive the simplest chemical formula for chromium oxide Cr_2O_3.

The simplest chemical formula for a substance consisting of three and more elements can be calculated by a similar technique.

Example 2. On complete burning of 2.66 g of a substance 1.54 g of carbon dioxide and 4.48 g of sulphurous anhydride were evolved. Determine the formula of the substance.

Solution. The composition of the combustion products indicates that the original substance consisted of carbon and sulphur. In addition to these elements, oxygen might have been a part of it as well.

To derive the simplest formula of the original substance one should determine the ratio of the masses of its components.

Since all carbon of the burned substance entered the carbon dioxide, its content of the burned 2.66 g of the original substance is equal to the weight in grams of carbon contained in 1.54 g of the carbon dioxide.

A gram-molecule of carbon dioxide weighs 44 g and contains 12 g of carbon. The quantity of carbon (x) contained in 1.54 g of carbon dioxide can be found from the proportion:

$$44:12 = 1.54:x$$

$$x = \frac{12 \times 1.54}{44} = 0.42 \text{ g}$$

By employing the same calculating technique the quantity of sulphur contained in 4.48 g of sulphurous anhydride can be determined. It is 2.24 g.

Thus we established that the burned substance contained 0.42 g of carbon per 2.24 g of sulphur. Since the sum of the two masses is equal to the mass of the original substance ($0.42 + 2.24 = 2.66$), it follows that the original substance contained no oxygen.

Following the model of the first example and using the obtained data one can calculate the ratio of the carbon atoms (x) to the sulphur atoms (y) in a molecule of the burned substance:

$$x:y = \frac{0.42}{12} : \frac{2.24}{32} = 0.035 : 0.070 = 1:2$$

The simplest formula of the burned substance is CS_2.

2. Derivation of the True Formula

To determine the true chemical formula of a compound, in addition to the composition of the substance, one should also know its molecular weight.

Example. An analysis of acetic acid shows that it contains 2.1 parts by weight of carbon per 0.35 part by weight of hydrogen and 2.8 parts by weight of oxygen. The density of acetic acid vapour with respect to hydrogen is 30. Hence its molecular weight is $30 \times 2 = 60$. What is the molecular formula of acetic acid?

Solution. Like in the previous case, we first determine the ratio of the number of carbon atoms (x) to atoms of hydrogen (y) and oxygen (z) in a molecule of acetic acid ($C_xH_yO_z$):

$$x:y:z = \frac{2.1}{12} : \frac{0.35}{1} : \frac{2.8}{16} = 0.175 : 0.35 : 0.175$$

By dividing all three terms of the last part of the equation by 0.175 we obtain:

$$x:y:z = 1:2:1$$

The simplest formula of acetic acid is thus CH_2O.

Now let us determine, what the molecular weight of acetic acid would be if this simplest formula corresponded to the true composition of its molecule. The calculation gives 30. But the molecular weight of acetic acid (found experimentally) is 60, that is two times as great. It follows, therefore, that the number of atoms in a molecule of acetic acid is twice as great as in the simplest formula. Hence the true chemical formula of acetic acid is $C_2H_4O_2$.

The true formulas of many gaseous substances reacting with other gases can also be determined by measuring the volumes and determining the composition of the gaseous reaction products (see Chapter II, Sec. 3).

The empirical formula very often corresponds to the true formula of a substance. For example, the simplest formula of carbon sulphide CS_2 corresponds to its molecular weight found experimentally (76), and hence it expresses the true chemical composition of its molecules.

PROBLEMS

153. Determine the formula of vanadium oxide knowing that its 2.73 g contain 1.53 g of the metal.

154. Determine the empirical formula of a substance containing 63.64 per cent of nitrogen and 36.36 per cent of oxygen.

155. Calculate the empirical formula of a substance consisting of hydrogen, carbon, oxygen and nitrogen combined in the weight ratio of 1 : 3 : 4 : 7.

156. Calculate the empirical formula of a substance containing 43.4 per cent of sodium, 11.3 per cent of carbon and 45.3 per cent of oxygen.

157. Calculate the formula of calcium chloride crystal hydrate knowing that its 7.3 g lose 3.6 g of water on dehydration.

158. Calculate the formula of sodium carbonate crystal hydrate knowing that on calcining its 14.3 g, 5.3 g of anhydrous sodium carbonate Na_2CO_3 are obtained.

159. Calculate the formula of a crystal hydrate containing 9.8 per cent of magnesium, 13 per cent of sulphur, 26 per cent of oxygen and 51.2 per cent of water.

160. A crystal hydrate contains 18.6 per cent of sodium, 25.8 per cent of sulphur, 19.4 per cent of oxygen and 36.2 per cent of water. Calculate the formula of the crystal hydrate.

161. What is the empirical formula of a silicate containing 40 per cent of MgO and 60 per cent of silica SiO_2?

162. Determine the formula of a mineral containing 73 per cent of zinc oxide ZnO and 27 per cent of silica SiO_2.

163. Calculate the formula of a substance containing 93.75 per cent of carbon and 6.25 per cent of hydrogen, if the density of its vapour with reference to air is 4.41.

164. Calculate the molecular formula of a substance in which 1 part by weight of hydrogen is combined with 6 parts by weight of carbon and 8 parts by weight of oxygen. The molecular weight of the substance is 180.

165. Calculate the molecular formula of a substance containing 84.2 per cent of sulphur and 15.8 per cent of carbon. The density of its vapour with reference to air is 2.62.

166. Calculate the molecular formula of butyric acid containing 54.5 per cent of carbon, 36.4 per cent of oxygen and 9.1 per cent of hydrogen knowing that the density of the acid vapour with respect to hydrogen is 44.

167. Calculate the molecular formula of aniline containing 77.4 per cent of carbon, 7.5 per cent of hydrogen and 15.1 per cent of nitrogen. The density of the aniline vapour with respect to air is 3.21.

168. On burning an organic substance consisting of carbon, hydrogen and sulphur, 2.64 g of carbon dioxide, 1.62 g of water and 1.92 g of sulphurous anhydride were obtained. Calculate the formula of the substance.

169. On burning 4.3 g of a hydrocarbon, 13.2 g of carbon dioxide were evolved. The density of the hydrocarbon vapour with respect to hydrogen is 43. Calculate the molecular formula of the hydrocarbon.

170. On burning 6.2 g of hydrosilicon, 12 g of silica SiO_2 were formed. The density of hydrosilicon vapour with respect to air is 2.14. Calculate the molecular formula of the hydro-silicon.

171. On complete burning of 13.8 g of an organic compound, 26.4 g of carbon dioxide and 16.2 g of water were formed. The density of vapour of this substance with respect to hydrogen is 23. Calculate the molecular formula of the organic substance.

172. For the purpose of determining the formula of a gaseous hydrocarbon, its 5 ml were mixed with 12 ml of oxygen and the mixture was exploded in a eudiometer*. On condensation of the water vapour, the volume of the gaseous residue (carbon dioxide and excess oxygen) was 7 ml, and after treatment with alkali to absorb carbon dioxide it reduced to 2 ml. All measurements were made in the same conditions. Calculate the formula of the hydrocarbon.

173. On explosion of a mixture consisting of one volume of a certain gas and two volumes of oxygen, two volumes of carbon dioxide and one volume of nitrogen were formed. Calculate the formula of the gas.

---

* A eudiometer is a graduated glass tube sealed on one end and immersed into mercury with its other end.

CHAPTER V

VALENCE.
DETERMINING THE FORMULA
FROM VALENCE

The valence of an element is the power of its atoms to combine with (or displace in molecules of various compounds) a definite number of atoms of other elements. The measure of valence is the number of hydrogen atoms with which an atom of a given element can unite or for which it can be substituted. An atom of a univalent element always combines with or can be substituted for only one atom of another univalent element. An atom of a bivalent element can combine with or be substituted for two atoms of a univalent element or one atom of another bivalent element, etc. Therefore the valence of an element can be judged by the composition of its compound with hydrogen as well as with any other element whose valence is known. For example, in the compound Na_2S sulphur is bivalent since its atom is bonded with two atoms of the univalent element sodium.

Finally, the valence of an element in a given compound can be determined from the relationship between the atomic weight of the element, its equivalent weight and the valence:

$$\text{valence} = \frac{\text{atomic weight}}{\text{equivalent weight}}$$

Chemical formulas of compounds consisting of two elements can be quickly determined if the valence of elements is known.

In the simplest case, where one of the component elements is univalent, the valence of the other element indicates directly the number of atoms of the first element per atom of the second in the molecule. For example, if aluminium is known to be trivalent, and chlorine in compounds with metals is always univalent, one can write at once the formula of aluminium chloride, $AlCl_3$.

In more complicate cases, to determine the formula of a compound, it is necessary so to select the number of atoms for each element that the product of the number of atoms of one element and its valence should equal the product of the number of atoms of the other element and its valence. In other words, the least common multiple for the valences of both elements should be found. The quotient of the division of the least common multiple by the valence of each element is the number of its atoms in the molecule of a given compound.

Example 1. Calculate the formula of perchloric anhydride knowing that the chlorine in this compound is heptavalent and oxygen is bivalent.

Solution. First determine the least common multiple of 2 and 7. It is 14. Divide 14 by 7 to obtain the number of atoms of chlorine in a molecule of perchloric anhydride. 14 divided by 2 is 7 which is the number of the oxygen atoms.

The formula of perchloric anhydride is Cl_2O_7.

Example 2. Derive the formula of bismuth sulphide knowing that bismuth is trivalent, and sulphur in compounds with metals is bivalent.

Solution. Since the least multiple of 2 and 3 is 6, the number of bismuth atoms in a molecule of its sulphide should be 3, and the number of sulphur atoms, 2.

The formula of bismuth sulphide is Bi_2S_3.

PROBLEMS

174. Determine the valence of manganese in MnS, Mn_2O_3, MnO_2, Mn_2O_7 and the valence of chromium in Cr_2S_3 and CrO_3.

175. What is the valence of nickel in its oxide if the equivalent weight of the metal in this compound is 19.57? (Atomic weight of nickel is 58.71.)

176. Calculate the valence of the metals in these compounds:

Li_2O, SnO_2, Mn_2O_7, CdS, $PbCl_2$, $CrCl_3$

177. Determine the valence of chlorine in the acids:

$HClO$, $HClO_2$, $HClO_3$, $HClO_4$

knowing that the valence of chlorine in them is the same as in the corresponding anhydrides.

54   *Problems in General Chemistry*

178. Determine the valence of sulphur in H_2SO_4, the valence of manganese in $HMnO_4$ and of silicon in H_2SiO_3, knowing that their valence is the same as in the corresponding anhydrides.

179. Calculate the valence of iodine in a compound containing 25.4 g of iodine, 0.2 g of hydrogen and 12.8 g of oxygen.

180. The chloride of a metal contains 69 per cent of chlorine. The atomic weight of the metal is 47.9. Determine its valence in the compound.

181. Calculate the equivalent weights of metals in Ag_2O, $Fe_2(SO_4)_3$, $FeCl_2$, V_2S_5, and $Cd_3(PO_4)_2$.

182. Determine the equivalent weight of chlorine in a compound containing 38.8 per cent of chlorine and 61.2 per cent of oxygen. Calculate the valence of chlorine from the found equivalent weight.

183. Calculate the formulas of the oxides of the elements given below (the valence is indicated by the Roman numerals):

I	II	III	IV	V	VI	VII	VIII
Ag	Ba	Al	Ti	Nb	Mo	I	Os

184. By making use of the Periodic System, calculate the formulas of highest oxides of vanadium, chromium and manganese.

185. Calculate the formulas of chlorides and sulphides of bivalent manganese, tetravalent tin and pentavalent antimony.

CHAPTER VI

CLASSIFICATION
OF INORGANIC COMPOUNDS

Inorganic compounds would be usually divided into five main classes, viz., oxides, acids, bases, amphoteric hydroxides and salts.

1. Oxides

Binary compounds composed of oxygen and one other element are called *oxides*, in which oxygen has a valence of 2 with respect to the other element. At the same time many elements in oxides have variable valence toward oxygen, and as a result, there may be several oxides for a given element. For example nitrogen forms N_2O, NO, N_2O_4, N_2O_5. Oxides where one atom of a given element combines with two or more atoms of oxygen are called *dioxides*, *trioxides*, etc.

The greater part of oxides combines, directly or indirectly, with water to form *hydroxides*. When heated, almost all hydroxides are decomposed into oxides and water.

Depending on their properties, hydroxides can be referred to the class of acids or bases. Besides, there are hydroxides having the properties of both acids and alkalis. These are *amphoteric hydroxides*. The oxides forming hydrates are also divided into three groups, namely acid oxides, basic oxides and amphoteric oxides.

Acid oxides. These are oxides whose hydrates are acids. They are mainly oxides of nonmetals, although some highest oxides of metals are also acid oxides (for example, CrO_3, Mn_2O_7 and others). Many acid oxides directly combine with water to form acids, for example

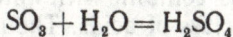

$$SO_3 + H_2O = H_2SO_4$$

Hydrates of other acid oxides are obtained indirectly.

Acid oxides are usually called *anhydrides* of the corresponding acids, for example phosphoric acid anhydride or simply phosphoric anhydride (P_2O_5).

The main feature of acid oxides is their power to react with alkalis to form salts, for example:

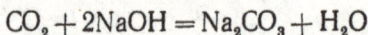

$$CO_2 + 2NaOH = Na_2CO_3 + H_2O$$

Acids as a rule do not react with acid oxides.

Basic oxides. This group of compounds contains oxides whose hydrates are bases. Basic oxides can only be formed by metals, and only oxides of the most active metals, like potassium, sodium, calcium and others, can directly combine with water to form water-soluble bases, alkalis. The majority of basic oxides do not react with water; the corresponding bases are formed indirectly and are insoluble in water.

All basic oxides react with acids to form salts, for example:

$$MgO + H_2SO_4 = MgSO_4 + H_2O$$

but do not react with alkalis.

Amphoteric oxides. These are intermediate oxides having the properties of both acid and basic oxides. They react with acids as well as with alkalis to form salts.

Since all oxides of the above discussed groups are capable of producing salts, they are often referred to as *salt-forming oxides*.

In addition to the salt-forming oxides, there are a few oxides which produce hydroxides neither directly nor indirectly. They do not react with acids, nor do they react with alkalis. These are *indifferent oxides*, for example, nitrogen oxide NO.

Peroxides form a special group. These are compounds of certain metals with oxygen, which can be referred to the class of oxides only on the formal grounds (by their composition), whereas in their essence they are salts of hydrogen peroxide, for example, sodium peroxide Na_2O_2, barium peroxide BaO_2, etc.

2. Acids

Oxides of nonmetals combine with water to form substances which belong to the class of acids. The majority of acids are hydroxides and therefore, in addition to hydrogen, they also contain oxygen. Hence the name *oxygen acids* (for example, H_2SO_4), to distinguish them from those having no oxygen, that is *anhydrous acids* like HCl and others.

Water solutions of acids taste sour and colour litmus red.

By the number of hydrogen atoms in a molecule that can be replaced by metals, acids can be distinguished as *monobasic* (for example, HNO_3), *dibasic* (for example, H_2SO_4), *tribasic* (H_3PO_4), etc.

All acids react with alkalis to form salts, for example:

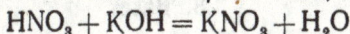

$$HNO_3 + KOH = KNO_3 + H_2O$$

If we subtract one or several hydrogen atoms capable of being replaced by metals, a group of atoms (sometimes only one atom) will remain which are called the *acid radicals*, and which participate in chemical reactions as a whole unit. For example, sulphuric acid H_2SO_4 gives two acid radicals viz., HSO_4 and SO_4, phosphoric acid H_3PO_4 gives three acid radicals, H_2PO_4, HPO_4, and PO_4. On these grounds molecules of all acids may be regarded as consisting of hydrogen atoms and acid radicals.

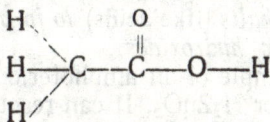

In the above acids, all hydrogen atoms can be replaced by metals, and their basicity is therefore determined by the number of hydrogen atoms in the molecule. But there are acids in which only a part of the hydrogen atoms can be replaced by metal. For example, only one out of four hydrogen atoms in acetic acid CH_3COOH can be replaced by metal. Acetic acid is therefore a monobasic acid although it contains four hydrogen atoms in its molecule.

The said difference in the properties of various acids is explained by the different structure of their molecules. It appears that *only hydrogen atoms bonded with oxygen atoms can be replaced by metals in the molecules of oxygen acids*.

The structural formulas of phosphoric acid in which all hydrogen atoms can be replaced by metal and acetic acid in which only one out of four hydrogen atoms can be replaced by metal explain this rule:

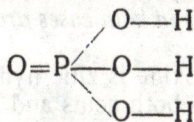

Phosphoric acid   *Acetic acid*

As evident from the formulas, all hydrogen atoms are connected with the oxygen atoms in a molecule of phosphoric acid, while in a molecule of acetic acid only one hydrogen atom is bonded with the oxygen atom and the other three are bonded with carbon. It follows therefore that only one atom of hydrogen can be replaced by metal in a molecule of acetic acid.

3. Bases

Bases are hydroxides which react with acids to form salts (but do not react with alkalis).

A molecule of any base contains an atom of a metal and one or several hydroxyl groups OH, for example, NaOH, $Mg(OH)_2$, $Bi(OH)_3$. When an acid acts on a base, hydroxyl groups are replaced by acid radicals and a salt is thus obtained, for example:

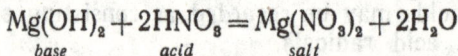

$$\underset{base}{Mg(OH)_2} + \underset{acid}{2HNO_3} = \underset{salt}{Mg(NO_3)_2} + 2H_2O$$

The majority of bases are insoluble in water. Bases formed by the most active metals like NaOH, $Ca(OH)_2$ are only soluble. These bases are known as *alkalis*. Their solutions have a soapy taste, colour litmus blue, and phenolphthalein crimson.

Atoms or groups of atoms which remain in a molecule of a base after subtraction of one, two or more hydroxyl groups are called the *basic radicals* by analogy with acid radicals. For example, by subtracting one after another hydroxyl groups from a molecule of copper hydroxide $Cu(OH)_2$ we get two basic radicals, viz., CuOH and Cu.

4. Amphoteric Hydroxides

Hydroxides that can react with acids (like bases) *as well as with alkalis* (like acids) *to form salts in both cases are called amphoteric hydroxides.*

An example of an amphoteric hydroxide is zinc hydroxide $Zn(OH)_2$ or H_2ZnO_2. It can react with both acids and alkalis according to the following equations:

$$Zn(OH)_2 + 2HCl = ZnCl_2 + 2H_2O$$
chloride

$$H_2ZnO_2 + 2NaOH = Na_2ZnO_2 + 2H_2O$$
zincate

In the first reaction zinc hydroxide performs the function of a base, and in the second it acts as an acid.

5. Salts

Salts are substances that may be regarded as products of hydrogen replacement in acids by basic radicals (metal atoms in particular) or as products of hydroxyl replacement in bases by acid radicals. In other words, molecules of all salts consist of acid and basic radicals.

The majority of salts are solid crystalline substances.

The following main types of salts are distinguished:

Normal or *neutral salts*, products of complete replacement of hydrogen atoms in acid molecules by metal atoms, or hydroxyl groups in base molecules by acid radicals, for example, Na_3PO_4, $Cu(NO_3)_2$.

Acid salts, products of incomplete replacement of hydrogen in acid molecules by metal atoms, for example, $NaHSO_4$, KH_2PO_4. Acid salts are obtained through the interaction of acids with bases in cases where the quantity of a base is insufficient to form a neutral salt:

$$H_2SO_4 + NaOH = NaHSO_4 + H_2O$$

Thus, acid salts still contain hydrogen capable of being replaced by metals. By acting with a sufficient amount of the corresponding base on an acid salt, one can replace the hydrogen by the metal and obtain a neutral salt:

$$NaHSO_4 + NaOH = Na_2SO_4 + H_2O$$

Basic salts are the products of incomplete replacement of hydroxyl groups in bases by acid radicals, for example, $Cu(OH)Cl$, $Bi(OH)_2NO_3$. Basic salts are obtained in cases where the amount of an acid taken to react with a base is insufficient to form a neutral salt. For example, if in the reaction between sulphuric acid and ferric hydroxide, one mole of $Fe(OH)_3$ is taken per one mole of H_2SO_4, a basic salt can be

formed:
$$Fe(OH)_3 + H_2SO_4 = FeOHSO_4 + 2H_2O$$

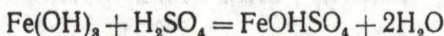

To convert a basic salt into a neutral one of the same acid, the basic salt should be acted upon with an amount of the acid sufficient to replace all hydroxyls by the acid radicals.

6. Deducing Formulas of Bases and Salts

As noted above, molecules of bases consist of metal atoms and hydroxyl groups, while molecules of salts of acid and basic radicals, which combine with each other in various definite proportions in each particular substance. Both hydroxyl group and acid and basic radicals have a definite valence whose value is determined as the valence of individual atoms.

The valence of the hydroxyl OH is 1, since it can combine with only one hydrogen atom to form a molecule of water.

The valence of an acid radical is determined by the number of hydrogen atoms which should be subtracted from a molecule of the acid in order to obtain the given radical. For example, phosphoric acid H_3PO_4 can produce three acid radicals, viz., monovalent $-H_2PO_4$, bivalent $>HPO_4$ and trivalent $\geqslant PO_4$.

The valence of a basic radical is determined by the number of hydroxyl groups subtracted from the base molecule. For example, calcium hydroxide $Ca(OH)_2$ forms two basic radicals, namely, monovalent radical $-CaOH$ and bivalent radical $>Ca$.

By extending the concept of valence to the above groups of atoms (the so-called radicals) one can make out quickly and easily formulas of any salt or base.

Example 1. Derive the formula of ferric hydroxide.

Solution. Since the hydroxyl is monovalent, a molecule of ferric hydroxide should contain one atom of iron and three hydroxyls.

The formula of the compound is $Fe(OH)_3$.

Example 2. Derive the formula of neutral calcium salt of phosphoric acid.

Solution. A molecule of this salt should contain bivalent atoms of Ca and trivalent acid radicals $\geqslant PO_4$. The least multiple of 2 and 3 is 6. Hence a molecule of the salt should contain $6 : 2 = 3$ atoms of calcium and $6 : 3 = 2$ acid radicals $>PO_4$.

The formula of the salt is $Ca_3(PO_4)_2$.

Example 3. Derive the formula of basic aluminium salt of hydrochloric acid.

Solution. The sought salt should contain monovalent acid radicals of hydrochloric acid —Cl and basic radicals of aluminium hydroxide comprising hydroxyl groups. Since aluminium hydroxide can produce two such radicals, viz., monovalent —$Al(OH)_2$ and bivalent $>AlOH$, it follows that two basic salts are possible: $Al(OH)_2Cl$ and $AlOHCl_2$.

Example 4. Derive the formula of acid magnesium salt of carbonic acid.

Solution. The salt should contain magnesium and a monovalent radical of carbonic acid —HCO_3. Since magnesium is bivalent, its one atom should combine with two acid radicals —HCO_3.

The formula of the salt is $Mg(HCO_3)_2$.

7. Most Important Methods for Preparation of Acids, Bases and Salts

Acids can be obtained by:

1. Direct combining of acid oxides with water, for example:

$$SO_3 + H_2O = H_2SO_4 \qquad (1)$$

2. Interaction between an acid and a salt (as a rule, by sulphuric acid reacting with a salt of the desired acid), for example:

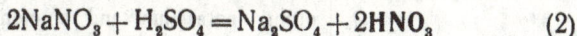

$$2NaNO_3 + H_2SO_4 = Na_2SO_4 + 2HNO_3 \qquad (2)$$

Preparation of bases:

1. Soluble bases (alkalis) can be produced by the reaction between the most active metals or their oxides with water:

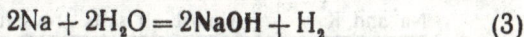

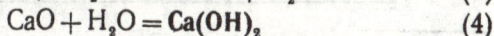

$$2Na + 2H_2O = 2NaOH + H_2 \qquad (3)$$
$$CaO + H_2O = Ca(OH)_2 \qquad (4)$$

2. Insoluble bases can be produced by the action of alkalis

on soluble salts of the corresponding metals, for example:
$$CuSO_4 + 2NaOH = \downarrow Cu(OH)_2^* + Na_2SO_4 \qquad (5)$$
Salts can be obtained:

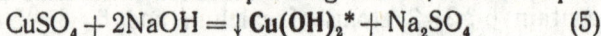

1. In reactions between an acid and a base (or amphoteric hydroxide) which neutralize each other to yield salt and water (*neutralization reaction*). For example:
$$HCl + NaOH = NaCl + H_2O \qquad (6)$$

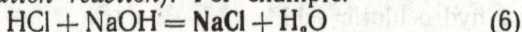

2. In reactions between an acid and a salt. A new acid and a new salt are produced as a result (see reaction (2)).

3. In reactions between an alkali and a *soluble* salt of metal forming an insoluble base. A new salt and an insoluble base are produced (see reaction (5)).

4. In reactions between two *soluble* salts. Two new salts are produced as a result:
$$NaCl + AgNO_3 = \downarrow AgCl + NaNO_3 \qquad (7)$$

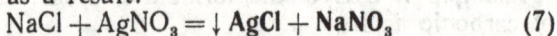

Solubility of Salts of Most Important Acids and Metals in Water

	Solubility of salts
Acids	
HNO_3	All salts are soluble
HCl	All salts are soluble except AgCl, CuCl, $PbCl_2$ and Hg_2Cl_2
H_2SO_4	All salts are soluble except $BaSO_4$, $SrSO_4$ and $PbSO_4$. Sparingly soluble are $CaSO_4$ and Ag_2SO_4
H_2CO_3	Of neutral salts only salts of sodium, potassium and ammonium are soluble
H_3PO_4	Ditto
H_2S	Ditto
Metals	
Na and K	Almost all salts are soluble

* An arrow on the left side of the formula shows that copper hydroxide falls out as precipitate.

The overwhelming majority of reactions between acids, bases and salts are exchange reactions. One of their products is always a new salt.

Practically, with the aid of these reactions, one can obtain new salts only on the condition that one of the reaction products is either insoluble or easily volatilized on heating and can thus be separated.

The following reactions are also very important for the production of salts:

5. Reaction between metals and acids, for example:

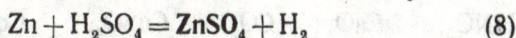

$$Zn + H_2SO_4 = ZnSO_4 + H_2 \qquad (8)$$

6. Reaction of basic and amphoteric oxides with acids, for example:

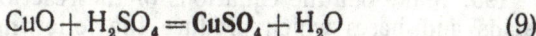

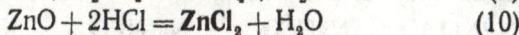

$$CuO + H_2SO_4 = CuSO_4 + H_2O \qquad (9)$$

$$ZnO + 2HCl = ZnCl_2 + H_2O \qquad (10)$$

PROBLEMS

186. Write formulas of oxides that can be obtained through decomposition by heating of the following hydroxides:

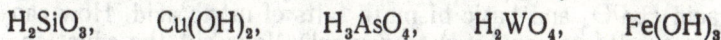

H_2SiO_3, $\qquad Cu(OH)_2$, $\qquad H_3AsO_4$, $\qquad H_2WO_4$, $\qquad Fe(OH)_3$

187. Determine formulas of anhydrides of the acids:

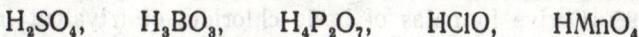

H_2SO_4, $\qquad H_3BO_3$, $\qquad H_4P_2O_7$, $\qquad HClO$, $\qquad HMnO_4$

188. Write acid radicals of the acid salts listed below:

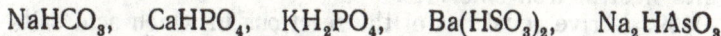

$NaHCO_3$, $\quad CaHPO_4$, $\quad KH_2PO_4$, $\quad Ba(HSO_3)_2$, $\quad Na_2HAsO_3$

What is their valence?

189. Write basic radicals that can be obtained from the following bases and amphoteric hydroxides:

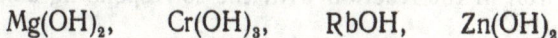

$$Mg(OH)_2, \qquad Cr(OH)_3, \qquad RbOH, \qquad Zn(OH)_2$$

What is their valence?

190. Derive formulas of neutral and acid salts of carbonic acid H_2CO_3 and arsenous acid H_3AsO_3 with potassium and calcium.

191. Derive formulas of acid potassium salts of arsenous acid H_3AsO_3, basic aluminium salts of monobasic acetic acid CH_3COOH, basic zinc salt of carbonic acid.

192. What substances can be produced in the reaction of an acid with a salt? acid with base? salt with salt? base with salt? Exemplify the reactions.

193. Which of the substances given below will react with hydrochloric acid?

N_2O_5, $Zn(OH)_2$, CaO, $AgNO_3$, H_3PO_4, H_2SO_4

Make out the equations of the reactions.

194. Which of the substances given below will react with sodium hydroxide?

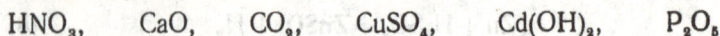

HNO_3, CaO, CO_2, $CuSO_4$, $Cd(OH)_2$, P_2O_5

Write the corresponding equations.

195. Make out the equations of the reactions between the acids and bases which produce the following salts:

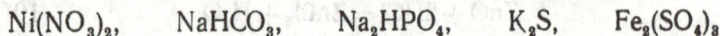

$Ni(NO_3)_2$, $NaHCO_3$, Na_2HPO_4, K_2S, $Fe_2(SO_4)_3$

196. Make out the equations of the reactions for producing magnesium chloride (a) by the action of acid on metal, (b) by the action of acid on base and (c) salt on salt.

197. Write formulas of acid calcium salts of phosphoric acid H_3PO_4 and basic bismuth salts of nitric acid. How can these salts be converted to neutral? Make out the equations of these reactions.

198. Derive formulas of basic chlorides of trivalent iron and the equations of the reactions in which these salts convert into neutral iron chlorides.

199. Derive equations of the reactions in which acid salts of potassium and calcium with sulphurous acid are obtained.

200. Derive the equations of the reactions for the formation of (a) basic magnesium chloride and (b) basic sulphate of trivalent iron in the reaction with the corresponding acids and bases.

201. How can ferric hydroxide be obtained from ferric chloride $FeCl_3$? orthophosphoric acid from its calcium salt $Ca_3(PO_4)_2$? cupric chloride $CuCl_2$ from copper oxide CuO?

202. By what reactions can one obtain nickelous oxide NiO and nickelous chloride $NiCl_2$, taking nickelous sulphate

---

* In solving problems marked with an asterisk consult the table of solubility of salts given on page 62.

$NiSO_4$ as the starting product and bearing in mind that both salts are soluble in water?

*203. Derive the equations of the reactions by which the following conversions can be effected:

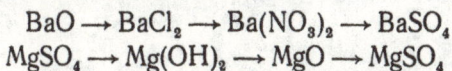

$$BaO \rightarrow BaCl_2 \rightarrow Ba(NO_3)_2 \rightarrow BaSO_4$$
$$MgSO_4 \rightarrow Mg(OH)_2 \rightarrow MgO \rightarrow MgSO_4$$

*204. How can this cycle of conversions be realized?

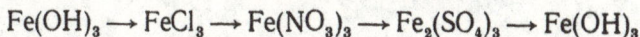

$$Fe(OH)_3 \rightarrow FeCl_3 \rightarrow Fe(NO_3)_3 \rightarrow Fe_2(SO_4)_3 \rightarrow Fe(OH)_3$$

*205. Indicate all reactions by which nitric acid can be obtained practically (bearing in mind its possible separation from other products of the reaction), if the following substances are available:

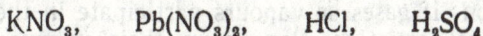

$$KNO_3, \qquad Pb(NO_3)_2, \qquad HCl, \qquad H_2SO_4$$

*206. Available are the salts K_2CO_3, $BaCl_2$, $NaCl$ and K_2SO_4. Which of them taken in pairs can be used to prepare potassium chloride KCl? Make out the equations of the reactions and prove your answer.

*207. It is possible to prepare a solution that would contain simultaneously $Ba(OH)_2$ and HCl, $CaCl_2$ and Na_2CO_3, $NaCl$ and $AgNO_3$, KCl and $NaNO_3$?

Which combinations are impossible and why?

*208. Is it possible to realize in solutions the following reactions:

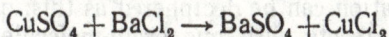

$$CuSO_4 + BaCl_2 \rightarrow BaSO_4 + CuCl_2$$
$$FeS + K_2SO_4 \rightarrow FeSO_4 + K_2S$$
$$AgCl + KNO_3 \rightarrow AgNO_3 + KCl$$

Develop your reasons paying attention to solubility of salts.

209. What new salts can be obtained from $CuSO_4$, $AgNO_3$, K_3PO_4 and $BaCl_2$? Write the equations of the reactions and name the obtained salts.

CHAPTER VII

CALCULATIONS
FROM CHEMICAL EQUATIONS

The chemical equation is an abbreviated record of the chemical reaction with the aid of symbols and formulas.

Each chemical equation contains exact data on the weight proportions of the reacting substances and the products of the reaction. If gases or vapours participate in the reaction the chemical equation also shows their volume ratios.

Let us consider, for example, the equation of the reaction between magnesium and dilute sulphuric acid:

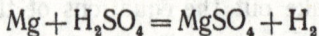

$$Mg + H_2SO_4 = MgSO_4 + H_2$$

The equation indicates that one atom of magnesium reacts with one molecule of sulphuric acid to form magnesium sulphate and hydrogen. Since the chemical formulas and symbols express not only the atoms or molecules but also weights of substances equal numerically to their atomic or molecular weights (for example, gram-atoms and gram-molecules), the above equation can be deciphered as this: one gram-atom of magnesium reacts with one gram-molecule of sulphuric acid to form one gram-molecule of magnesium sulphate and one gram-molecule of hydrogen. One gram-atom of magnesium weighs 24 g, gram-molecule of sulphuric acid 98 g, one gram-molecule of magnesium sulphate 120 g and gram-molecule of hydrogen weighs 2 g:

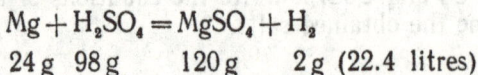

$$Mg + H_2SO_4 = MgSO_4 + H_2$$
$$24 g \quad 98 g \qquad 120 g \qquad 2 g \ (22.4 \ litres)$$

Numbers 24, 98, 120 and 2 express the weight relations between the masses of the reacting substances. In practical realization of the reaction, 98 g of sulphuric acid are required to react with 24 g of magnesium. As a result, 120 g of magnesium sulphate and 2 g of hydrogen (or 22.4 litres at 0°C and

760 mm Hg) are produced. If excess sulphuric acid is present in the reaction, it will remain unreacted, and if the acid is deficient, a part of magnesium will not react. As a result the quantities of magnesium sulphate and hydrogen obtained in the reaction will be less than calculated.

Since the chemical equation also reflects the quantitative aspect of the process, it can be used for various calculations, for example, for determining the quantities of the starting materials required to produce a given quantity of a substance, or the quantities of the reaction products that can be obtained upon reaction from a given quantity of the starting materials, etc. In solving such problems, the requisite condition is in the first instance the *correct chemical equation of the reaction*. Next the ratios of the reacting masses or volumes should be determined, and finally it is necessary to make out the requisite proportions from which the sought values could be calculated.

Example 1. How many grams of sulphuric acid are required to neutralize 20 g of sodium hydroxide?

Solution. Knowing that the products of the reaction between an acid and an alkali are always salt and water, we can derive the equation of the reaction, selecting the coefficients in such a way that the number of atoms of each element in the right and the left parts of the equation will be equal:

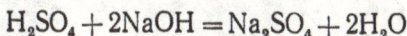

$$H_2SO_4 + 2NaOH = Na_2SO_4 + 2H_2O$$

From the equation it follows that in order to neutralize one gram-molecule of sulphuric acid, two gram-molecules of sodium hydroxide are required. By calculating the molecular weights of sulphuric acid and sodium hydroxide we find out that the gram-molecule of sulphuric acid weighs 98 g, and two gram-molecules of sodium hydroxide weigh 80 g. Hence the conditions of the problem: 98 g of sulphuric acid neutralize 80 g of sodium hydroxide; what quantity in grams of sulphuric acid is required to neutralize 20 g of sodium hydroxide? Let the sought quantity of sulphuric acid be x. Then, the condition of the problem can be expressed as this:

$$98\,g\ H_2SO_4\ \text{neutralize}\ 80\,g\ NaOH$$
$$x\,g\ H_2SO_4\ \text{neutralize}\ 20\,g\ NaOH$$

The proportion can thus be made

$$98:80 = x:20$$

whence

$$x = \frac{98 \times 20}{80} = 24.5 \, g$$

Example 2. How many litres of carbon dioxide are evolved on burning 13 litres of acetylene C_2H_2 if both gases are measured in the same conditions?

Solution. The equation of the reaction is

$$2C_2H_2 + 5O_2 = 4CO_2 + 2H_2O$$

Since the coefficients in the formulas indicate the relative volumes of the reacting gases, the problem can be read as this: When burned, 2 litres of acetylene form 4 litres of carbon dioxide. How many litres of carbon dioxide will be obtained on burning 13 litres of acetylene?

Schematically, the condition of the problem can be represented as follows:

2 litres of C_2H_2 form 4 litres of CO_2

13 litres of C_2H_2 form x litres of CO_2

Using the proportion method,

$$2:4 = 13:x$$

Solving for x,

$$x = \frac{4 \times 13}{2} = 26 \text{ litres}$$

Example 3. Chlorine can be obtained by the action of sulphuric acid and manganese dioxide on sodium chloride. The equation of the reaction is:

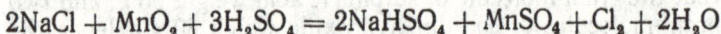

$$2NaCl + MnO_2 + 3H_2SO_4 = 2NaHSO_4 + MnSO_4 + Cl_2 + 2H_2O$$

How many litres of chlorine at 0°C and 760 mm Hg can be obtained from 100 g of sodium chloride?

Solution. The equation of the reaction shows that two gram-molecules of sodium chloride form one gram-molecule of chlorine. Having calculated the mass of two gram-molecules of sodium chloride and knowing that one gram-molecule of chlorine at 0°C and 760 mm Hg occupies a volume of 22.4

litres, one can write the condition of the problem as this:

117 g of NaCl produce 22.4 litres of Cl_2

100 g of NaCl produce x litres of Cl_2

The following proportion is made:

$$117 : 22.4 = 100 : x$$

Solving for x:

$$x = \frac{22.4 \times 100}{117} = 19.15 \text{ litres}$$

PROBLEMS

210. The reaction of ammonia burning is expressed by the equation:

$$4NH_3 + 3O_2 = 6H_2O + 2N_2$$

How do the masses and volumes of the reacting substances relate?

211. On dissolving in sulphuric acid, magnesium formed 36 g of manganese sulphate. What was the mass of the magnesium? How many grams of sulphuric acid were consumed for its dissolution?

212. To a solution containing 10 g of H_2SO_4 were added 9 g of NaOH. What is the reaction of the obtained solution: acid, neutral or alkaline?

213. When heated, calcium carbonate $CaCO_3$ decomposes into lime and carbon dioxide. What is the approximate weight of natural lime stone (containing 90 per cent of $CaCO_3$) required to obtain 7 tons of lime?

214. What weight in grams of sodium hydroxide is required to convert 100 g of copper sulphate into copper hydroxide?

215. How many grams of potassium hydroxide are required to convert 70 g of sulphuric acid into an acid salt? into a neutral salt?

216. How many gram-molecules of water will be liberated in the reduction of 200 g of copper oxide to metallic copper?

217. How many litres of hydrogen (at STP) will be evolved in the reaction between water and 15 g of calcium?

218. What quantities of zinc and 20 per cent sulphuric acid are required to liberate 56 litres of hydrogen at STP?

219. How many litres of acetylene (at 0°C and 760 mm Hg) can be obtained by the action of water on 8 g of calcium carbide according to the equation:

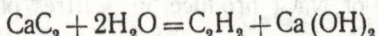

$$CaC_2 + 2H_2O = C_2H_2 + Ca(OH)_2$$

220. In laboratory conditions carbon dioxide is prepared by the action of hydrochloric acid on marble. Assuming that marble consists entirely of pure $CaCO_3$, what weight of marble is required to prepare 84 litres of carbon dioxide at STP?

221. How many litres of fire damp (at STP) will be liberated in decomposition of one gram-molecule of water by electric current?

222. What weight in grams of potassium chlorate $KClO_3$ is required to liberate 100 litres of oxygen at STP?

223. How many litres of oxygen at STP are required to burn 100 g of ethyl alcohol C_2H_5OH?

224. What quantity of sodium chloride is required to prepare 56 litres of hydrogen chloride at STP?

225. How many litres of oxygen at STP are required to reduce 120 g of molybdic anhydride MoO_3 to metallic molybdenum?

226. How many litres of carbon dioxide (at 0°C and 760 mm Hg) should be passed through a solution of calcium hydroxide $Ca(OH)_2$ to obtain 25 g of $CaCO_3$?

227. On passing water vapour over heated carbon, water gas is produced, which consists theoretically of equal volumes of carbon monoxide and hydrogen:

$$C + H_2O = CO + H_2$$

How many cubic metres of water gas measured at STP can be obtained from 3 kg of carbon?

228. Assuming that the oxygen content of air is 20 per cent by volume, calculate what volume of air is required to burn one cubic metre of technical water gas containing 50 per cent of hydrogen, 40 per cent of carbon monoxide, 5 per cent of carbon dioxide, and 5 per cent of nitrogen.

229. How many litres of air are required to burn one cubic metre of lighting gas containing 50 per cent of hydrogen, 35 per cent of methane, 8 per cent of carbon monoxide, 2 per cent of ethylene and 5 per cent of noncombustible admixtures? Assume the oxygen content of air to be 20 per cent by volume.

230. Mixed together are 7.3 g of hydrogen chloride and 4 g of ammonia. How many grams of ammonium chloride NH_4Cl will be formed? What gas and in what quantity remains unreacted?

231. A solution containing 34 g of silver nitrate $AgNO_3$ is mixed with a solution containing the same quantity of sodium chloride NaCl. Will all silver nitrate participate in the reaction? How many grams of silver chloride will be formed in the reaction?

232. Solutions containing equal masses of nitric acid and sodium hydroxide are mixed together. What is the reaction of the resulting solution: acid, alkaline or neutral?

233. In a mixture of 20 moles of hydrogen and 20 moles of nitrogen, 4 moles of ammonia were formed. How many moles of nitrogen and hydrogen remained in the mixture?

234. A mixture of 10 moles of sulphurous anhydride and 15 moles of oxygen were passed over a catalyst. 8 moles of sulphuric anhydride SO_3 formed as a result. How many moles of sulphurous anhydride and oxygen remained unreacted?

235. To a solution containing 0.2 mole of ferric chloride $FeCl_3$ 0.24 mole of sodium hydroxide NaOH was added. How many moles of ferric hydroxide $Fe(OH)_3$ were formed in the reaction and how many moles of ferric chloride $FeCl_3$ remained in the solution?

236. For preparing a certain quantity of cupric sulphate $CuSO_4$, 20 g of copper oxide CuO were heated in a solution containing 21 g of sulphuric acid. Has all copper oxide dissolved in the acid? How many grams of copper sulphate were formed in the reaction?

237. How can sodium carbonate Na_2CO_3 be converted to sodium chloride? How many grams of sodium chloride can be obtained from 265 g of soda?

238. What weights of soda Na_2CO_3 and calcium chloride $CaCl_2$ are required to obtain 200 g of calcium carbonate $CaCO_3$?

239. How can copper nitrate be prepared if blue vitriol, nitric acid and alkali are available? Write the equations of the reactions and calculate how many grams of $Cu(NO_3)_2 \cdot 3H_2O$ can be prepared from 450 g of 20 per cent nitric acid.

240. Write the chemical equations of the reactions in which copper oxide CuO can be obtained from copper chloride $CuCl_2$.

Calculate the weight of copper chloride required to prepare 32 g of copper oxide.

241. When burned, 3 g of anthracite liberated 5.3 litres of carbon dioxide measured at standard conditions of temperature and pressure. What was the carbon content of the anthracite?

242. To neutralize 50 g of a solution of sulphuric acid, 2 g of sodium hydroxide were spent. What was the percent concentration of the sulphuric acid?

243. On dissolution of 0.8 g of zinc dust (zinc powder containing a small quantity of zinc oxide admixture) in sulphuric acid, 224 ml of hydrogen were liberated as measured at STP. What was the zinc content (in per cent) of the zinc dust?

244. In order to determine sodium chloride content of technical sodium hydroxide, its 2 g were dissolved in water and a solution of silver nitrate $AgNO_3$ was added until the formation of silver chloride precipitate stopped. The washed and dried precipitate AgCl weighed 0.287 g. What was the percentage of sodium chloride in the sodium hydroxide?

245. Into a 1-litre flask containing 100 ml of 10 per cent solution of HCl at a temperature of 21°C were placed 3.25 g of zinc and the flask was stoppered at once. What will be the pressure inside the flask on termination of the reaction, if the temperature remains unchanged? The density of the acid can be assumed to be equal to 1, and the initial pressure 1 atm.

CHAPTER VIII

THERMOCHEMICAL EQUATIONS
AND CALCULATIONS

Any chemical reaction is accompanied by the evolution or absorption of energy, mostly in the form of heat. The amount of heat can be measured and expressed in any unit of energy (joules, calories, etc.).

Chemical equations in which the thermal effect of the reaction is specified are called *thermochemical equations*, in which the number of calories is always referred to gram-molecular quantities of the reacting substances or the reaction products. In exothermal reactions the amount of liberated heat is given with a plus sign $(+)$ and in the endothermal reactions a minus sign $(-)$ is employed. For example, the thermochemical equation of nitrogen oxide formation

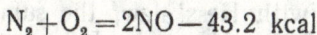

$$N_2 + O_2 = 2NO - 43.2 \text{ kcal}$$

shows that when two gram-molecules of nitrogen oxide are formed from nitrogen and oxygen, 43.2 kilocalories are absorbed.

Since in the thermochemical equations the symbols and formulas designate gram-atoms and gram-molecules of the reacting substances, fractional coefficients can be used*. Therefore, the previous equation can also be expressed as this:

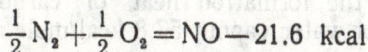

$$\frac{1}{2} N_2 + \frac{1}{2} O_2 = NO - 21.6 \text{ kcal}$$

The amount of heat evolved or absorbed during the formation of one gram-molecule of a chemical compound from simple substances is called the *heat of formation* of the compound. From the above equation it follows, for example, that the formation heat of nitrogen oxide is 21.6 kcal/mole.

---

* Fractional coefficients are used to indicate the thermal effect of the reaction as referred to one mole of the formed substance.

All thermochemical calculations are based on the following principles:

1. *The amount of heat evolved (absorbed) in decomposition of a compound into simple substances is the same as that absorbed (evolved) during formation of this chemical compound from the initial simple substances.*

It follows therefore that the heat of formation of a chemical compound is equal to the heat of its decomposition taken with opposite sign.

2. *The thermal effect of a chemical process is the sum of the thermal effects of all intermediate steps of the process* (**Hess' law** of heat summation).

For instance, the process of preparation of carbon dioxide from carbon and oxygen can be effected in one step:

$$C + O_2 = CO_2 + 94 \text{ kcal}$$

But the same process can also be realized in two steps, namely:

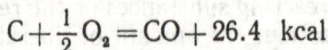

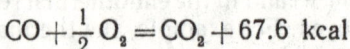

$$C + \frac{1}{2} O_2 = CO + 26.4 \text{ kcal}$$

$$CO + \frac{1}{2} O_2 = CO_2 + 67.6 \text{ kcal}$$

In accordance with Hess' law, the sum of the thermal effects of the two last reactions should be equal to the thermal effect of the first reaction, and this proves true $(26.4 + 67.6 = = 94 \text{ kcal})$.

Let us consider a few examples.

Example 1. Determine the thermal effect of the reaction of methane burning:

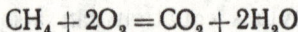

$$CH_4 + 2O_2 = CO_2 + 2H_2O$$

knowing that the formation heat of carbon dioxide is 94 kcal/mole, of water vapour 57.8 kcal/mole and of methane 17.9 kcal/mole.

Solution. Let us divide the reaction into steps to make the course of the reaction more vivid:

(1) Decomposition of methane into carbon and hydrogen

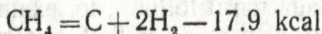

$$CH_4 = C + 2H_2 - 17.9 \text{ kcal}$$

(2) Formation of carbon dioxide from carbon and oxygen

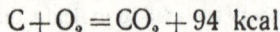

$$C + O_2 = CO_2 + 94 \text{ kcal}$$

(3) Formation of water vapour from hydrogen and oxygen

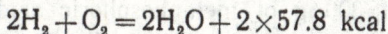

$$2H_2 + O_2 = 2H_2O + 2 \times 57.8 \text{ kcal}$$

The sum of the thermal effects of the three steps of the reaction should be equal to the total thermal effect of the reaction of methane burning:

$$-17.9 \text{ kcal} + 94 \text{ kcal} + 2 \times 57.8 \text{ kcal} = 191.7 \text{ kcal}$$

Hence the thermochemical equation of the reaction of methane burning will be

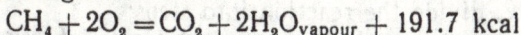

$$CH_4 + 2O_2 = CO_2 + 2H_2O_{vapour} + 191.7 \text{ kcal}$$

By expressing the effect of the reaction in units of heat (energy, work) adopted in the International System of Units, i.e. in joules (or kilojoules), and bearing in mind that 1 kcal equals 4,186.8 J or 4.1868 kJ, we can find that the thermal effect of the reaction is equal to $191.7 \times 4.1868 = 802$ kJ.

Example 2. Determine the formation heat of nitrous oxide from the equation

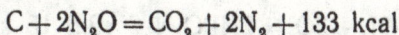

$$C + 2N_2O = CO_2 + 2N_2 + 133 \text{ kcal}$$

Solution. Split the reaction into steps:

(1) Decomposition of nitrous oxide into nitrogen and oxygen (designate the unknown heat effect of this reaction by x)

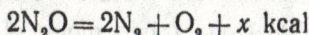

$$2N_2O = 2N_2 + O_2 + x \text{ kcal}$$

(2) Formation of carbon dioxide from carbon and oxygen

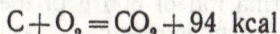

$$C + O_2 = CO_2 + 94 \text{ kcal}$$

The sum of the thermal effects of the last two reactions should be equal to the thermal effect of the first reaction:

$$x \text{ kcal} + 94 \text{ kcal} = 133 \text{ kcal}$$

whence

$$x = 133 - 94 = 39 \text{ kcal}$$

Thus, in decomposition of two moles of nitrous oxide, 39 kcal are evolved. The same amount of calories is absorbed during the formation of two moles of this substance. It follows therefore that the heat of formation of nitrous oxide is

$$\frac{-39}{2} = -19.5 \text{ kcal/mole}$$

or -81.64 kJ/mole.

Example 3. Calculate the amount of heat liberated during preparation of 300 g of metaphosphoric acid HPO_3 from phosphoric anhydride P_2O_5, basing on the following data: the formation heat of phosphoric anhydride is 360 kcal/mole, of metaphosphoric acid 221.15 kcal/mole, and of water 68.3 kcal/mole.

Solution. Let us derive the equation for the reaction by designating the unknown thermal effect through x:

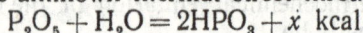

$$P_2O_5 + H_2O = 2HPO_3 + x \text{ kcal}$$

To find x, divide the reaction into steps:

(1) Decomposition of phosphoric anhydride into phosphorus and oxygen

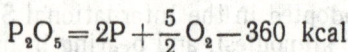

$$P_2O_5 = 2P + \frac{5}{2}O_2 - 360 \text{ kcal}$$

(2) Decomposition of water into hydrogen and oxygen

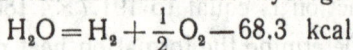

$$H_2O = H_2 + \frac{1}{2}O_2 - 68.3 \text{ kcal}$$

(3) Formation of metaphosphoric acid from the elements

$$2P + H_2 + 3O_2 = 2HPO_3 + 2 \times 221.15 \text{ kcal}$$

From the sum of the thermal effects of individual reactions we can find the total thermal effect of the reaction:

$$x = -360 - 68.3 + 442.3 = 14 \text{ kcal}$$

Hence in the formation of two moles of metaphosphoric acid (120 g), 14 kcal were evolved. The amount of heat (y) liberated during formation of 300 g of metaphosphoric acid can be found from the proportion:

$$120 : 300 = 14 : y$$

$$y = \frac{300 \times 14}{120} = 35 \text{ kcal, or } 14.65 \text{ kJ}$$

* * *

Many thermochemical calculations can be simplified if the following principle derived from Hess' law is applied:

The thermal effect of a chemical reaction equals the sum of the formation heats of the resulting substances minus the sum of the formation heats of the reactants.

For example, the thermal effect of the reaction

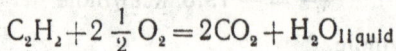

$$C_2H_2 + 2\frac{1}{2}O_2 = 2CO_2 + H_2O_{\text{liquid}}$$

is equal to

$$2 \times 94 \text{ kcal} + 68.3 \text{ kcal} - (-54.2 \text{ kcal}) = 310.5 \text{ kcal}$$

formation heat   formation   formation
of two moles   heat of   heat of
of CO_2 liquid   acetylene
        water

Heat of Formation of Some Compounds, kcal/mole

H_2O_{vapour}	57.8	NO	—21.6	CO_2	94.0
H_2O_{liquid}	68.3	P_2O_5	360	CuO	37.5
HCl	22.1	CH_4	17.9	CaO	151.7
SO_2	71.0	C_2H_4	—12.5	$Ca(OH)_2$	236.0
NH_3	11.0	C_2H_2	—54.2	Fe_2O_3	195.2
N_2O	—19.5	CO	26.4	Al_2O_3	393.3

PROBLEMS

246. During reaction of 2.1 g of iron with sulphur 0.855 kcal was evolved. Calculate the heat of formation of ferric sulphide.

247. In the determination of the formation heat of zinc oxide, 3.25 g of zinc were burned in a calorimetric bomb; 4.17 kcal were liberated. Calculate the heat of formation of zinc oxide.

248. During oxidation of 9 g of aluminium by oxygen, 65.55 kcal were evolved. Determine the heat of formation of aluminium oxide Al_2O_3.

249. Calculate the heat evolved in an explosion of 8.4 litres of fire damp taken at STP. The heat of formation of water vapour is 57.8 kcal/mole.

250. The reaction of formation of hydrogen chloride from chlorine and hydrogen is expressed by the equation

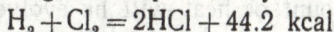

$$H_2 + Cl_2 = 2HCl + 44.2 \text{ kcal}$$

Calculate the heat evolved in the reaction between 1 litre of chlorine and hydrogen.

251. Calculate the heat evolved on burning one cubic metre of hydrogen taken at STP, if the reaction product is water.

252. Determine the thermal effect of the reaction of sulphur burning in nitrous oxide:

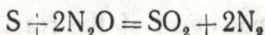

$$S + 2N_2O = SO_2 + 2N_2$$

253. Determine the heat of formation of hydrogen phosphide PH_3 from the equation

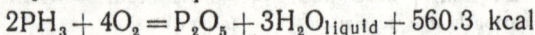

$$2PH_3 + 4O_2 = P_2O_5 + 3H_2O_{liquid} + 560.3 \text{ kcal}$$

254. Determine the heat of formation of hydrogen sulphide from the equation

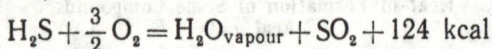

$$H_2S + \frac{3}{2}O_2 = H_2O_{vapour} + SO_2 + 124 \text{ kcal}$$

255. Determine the thermal effect of the reaction:

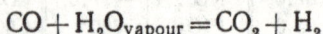

$$CO + H_2O_{vapour} = CO_2 + H_2$$

256. Determine the heat of formation of methyl alcohol CH_3OH from the reaction

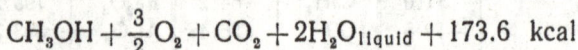

$$CH_3OH + \frac{3}{2}O_2 + CO_2 + 2H_2O_{liquid} + 173.6 \text{ kcal}$$

257. Determine the heat of formation of copper oxide knowing that during reduction of one gram-molecule of copper oxide with carbon (with formation of CO) 11.1 kcal are absorbed.

258. Determine the heat of formation of carbon sulphide CS_2 from the reaction

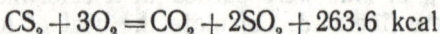

$$CS_2 + 3O_2 = CO_2 + 2SO_2 + 263.6 \text{ kcal}$$

259. Determine the amount of heat that will be evolved in burning 100 litres of ethylene C_2H_4 at STP, if the vapour is condensed to water.

260. The reaction of ethylene C_2H_4 burning is expressed by the equation

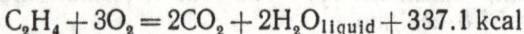

$$C_2H_4 + 3O_2 = 2CO_2 + 2H_2O_{liquid} + 337.1 \text{ kcal}$$

Determine the heat of formation of ethylene.

261. What amount of heat will be evolved on burning 112 litres of water gas (which is a mixture of equal volumes of hydrogen and carbon monoxide) if carbon dioxide and water vapour are formed as a result?

262. The main process in the blast furnace can be expressed by the summary equation:

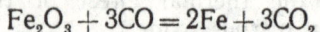

$$Fe_2O_3 + 3CO = 2Fe + 3CO_2$$

Determine the thermal effect of the reaction.

263. What amount of heat will be evolved during reduction of 8 g of copper oxide by hydrogen with the formation of liquid water?

264. The reaction of decomposition of calcium carbonate is expressed by the reaction:

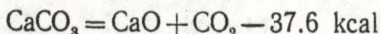

$$CaCO_3 = CaO + CO_2 - 37.6 \text{ kcal}$$

Determine the heat of formation of calcium carbonate from calcium, oxygen and carbon.

265. Determine the heat of formation of ammonium chloride from the reaction:

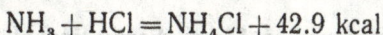

$$NH_3 + HCl = NH_4Cl + 42.9 \text{ kcal}$$

266. Equal volumes of hydrogen and acetylene taken at equal conditions were burned. In which case the amount of the evolved heat was greater? How much greater?

Assume the formation heat of water to be equal to 57.8 kcal/mole.

267. To reduce metals from their oxides, the following reductants are used: carbon (which is oxidized in the reaction to carbon monoxide), hydrogen, or carbon monoxide. Determine the thermal effects of Fe_2O_3 reduction with all three reductants.

CHAPTER IX

ATOMIC STRUCTURE.
CHEMICAL BONDING

1. Structure of Atom. Formation of Ions

An atom of each element consists of a nucleus bearing the positive charge and electrons which revolve around the nucleus and bear the negative charge. The mass of an electron is quite small, measuring only $1/_{1840}$ mass of a hydrogen atom. Therefore the entire mass of an atom is practically concentrated in its nucleus. The positive charge of a nucleus equals the sum of the negative charges of the electrons which revolve around it. The atom as a whole is therefore electrically neutral.

The charge of an electron is the least known quantity of electricity $(4.80 \times 10^{-10}$ electrostatic units). If we take this magnitude as a unit, the charge of an atom expressed in these units will be numerically equal to the atomic number of the element which shows the place occupied by the given element in the Periodic Table.

Electrons revolving around the nucleus are said to be at certain *energy levels*. The energy levels would be usually classified according to shells and subshells. Each subshell includes a certain fixed number of electrons, characterized by almost equal energy. The shells of the atom, which are thought to be concentric with the nucleus, are denoted by the letters K, L, M, N, O,..., the letter K denoting the shell closest to the nucleus and having the lowest energy level. Electrons of each subsequent shell are at a higher energy level. The greatest number of electrons that can be filled in a given shell (i.e. that can be at a given energy level) equals $2\,n^2$, where n is the number of the energy level. The number of electrons in the outer shell of all elements does not exceed eight.

The electrons of the outermost shell, which are the farthest from the nucleus and therefore the least strongly connected with it, can break away from the atom and be captured by other atoms, taking up the position in their outer shell. The loss or gain of electrons by an atom makes it electrically charged. In the former case the atom becomes positively charged and in the latter negatively charged. Thus formed charged particles are called *ions*.

Ions are conventionally denoted by the same symbol as the corresponding atom with a superscript added at the right consisting of the same number of plus or minus signs as there are units in the charge of the ion. For example, the positive doubly charged calcium ion is denoted by the symbol Ca^{++} or Ca^{2+}, the negative singly charged chloride ion by the symbol Cl^-, and so on.

The magnitude of the charge depends on the number of electrons the atom has lost or gained. For instance, if a magnesium atom having a total of 12 electrons in its shell loses two electrons from its outer shell, the resulting magnesium ion will have a charge of $+2$, since the loss of electrons does not alter the charge of the nucleus which is $+12$, but the total charge of the remaining electrons will now be -10 (the charge of the ion will be $-10 +12=+2$).

Here is another example. An atom of sulphur has a total of 16 electrons, six of which are in the outer shell. If it acquires two more electrons, a negative doubly charged sulphur ion will be formed, since the total charge of the electrons becomes -18, whereas the charge on the nucleus is $+16$ (the charge on the ion will be $-18 +16=-2$).

Thus, gaining a certain number of electrons converts an atom into an ion whose negative charge is equal numerically to the number of the electrons gained. And vice versa, the loss of a certain number of electrons turns the atom into an ion whose positive charge equals the number of the lost electrons.

Many ions in turn can lose or gain electrons and thus become either electrically neutral atoms, or ions with other charges.

When an ion loses electrons, its positive charge increases or its negative charge decreases, or else becomes zero (i.e. the ion becomes an electrically neutral atom). On the other hand, the addition of electrons to an ion decreases its positive or increases its negative charge.

Thus for example, if the positive doubly charged ferrous ion Fe^{2+} loses one electron it becomes a triply charged ferric ion Fe^{3+}, or if it gains two electrons it turns into an electrically neutral atom Fe. If a quarterly charged tin ion Sn^{4+} gains two electrons, it turns into a doubly charged ion Sn^{2+}. If it gains four electrons, an electrically neutral atom of tin is obtained. If the negative doubly charged sulphide ion S^{2-} loses two electrons, it turns into a sulphur atom S, and so on.

By denoting the electron by the symbol e^- we can express the above transformations as this:

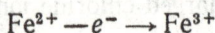

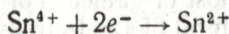

$$Fe^{2+} - e^- \longrightarrow Fe^{3+}$$
$$Fe^{2+} + 2e^- \longrightarrow Fe$$
$$Sn^{4+} + 2e^- \longrightarrow Sn^{2+}$$
$$Sn^{4+} + 4e^- \longrightarrow Sn$$
$$S^{2-} - 2e^- \longrightarrow S$$

The ability of an atom to turn into positively or negatively charged ions depends on the position of a particular element in the Periodic System of Mendeleyev. Atoms of the elements standing in the beginning of the period have the charge on the nucleus which is less than in the atoms of the elements in the end of the period. In the former case the electrons are attracted weaker than in the latter case. Therefore the ability of atoms to gain electrons increases in the periods from left to right. The number of electrons in the outer shell of the atom increases too. Only atoms whose outer shell contains more than five electrons (atoms of nonmetals) can turn into electronegative ions. Atoms having less than four electrons in the outer shell (except hydrogen atom) can lose their electrons only, and as far as we know, can never gain them. These are atoms of the elements which we call metals.

The loss or gain of electrons takes place in various chemical processes, in particular in the formation of many compounds out of simple substances, where atoms of the elements combine with each other.

2. Compounds with Electrovalent and Atomic Bonds

All substances, except metals, can be divided into two major groups by the character of chemical bonds existing between their atoms.

1. Substances having the so-called *electrovalent* or *ionic bonds* consist of positively and negatively charged ions attracted to each other by the electrostatic force. Compounds having such bonds are called *ionic*.

2. Substances having *covalent* or *atomic bonds* consist of electroneutral molecules formed by atoms and bonded by the interaction of electrons which are shared by the two joined atoms. Compounds having covalent bonds are called *atomic* or *covalent compounds*.

Salts, basic oxides and others are representatives of ionic compounds. Simple gases, like hydrogen, oxygen, nitrogen, are substances with atomic bonds.

It is impossible to draw a sharp boundary line between compounds belonging to these two major groups, since there are many substances that possess both ionic and atomic bonds.

3. Formation of Ionic Compounds

In accordance with the present-day theory, compounds with ionic bonds are formed as follows.

When two or more atoms, some of which tend to accept and the others to donate electrons, come together, the electrons in their outer shells rearrange. The resulting charged ions attract each other to form a compound with ionic bonds.

For example, in the formation of sodium chloride from the elements, sodium atoms which have only one electron in their outer shell approach chlorine atoms which contain seven electrons in the outer shell. The sodium atoms yield their electrons to the chlorine atoms and thus turn into positive singly charged sodium ions. At the same time the chlorine atoms which accepted one electron each into their outer shells become negative singly charged chloride ions. The electrostatic attractive force which arises between the unlike charges bonds the ions into the salt sodium chloride (Fig. 1).

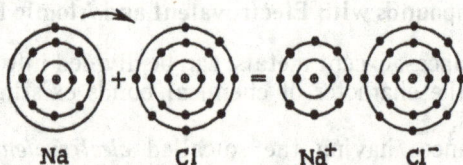

Fig. 1. Formation of sodium chloride

The processes taking place in the formation of sodium chloride can be expressed by the following *electronic equations*:

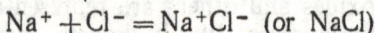

$$Na - e^- = Na^+$$
$$Cl + e^- = Cl^-$$
$$Na^+ + Cl^- = Na^+Cl^- \text{ (or NaCl)}$$

Likewise, during formation of aluminium oxide out of the elements, each aluminium atom donates three electrons to the atoms of oxygen. Each oxygen atom can accept two electrons only since its outer shell is already filled with six electrons. As a result, positive triply charged ions of aluminium Al^{3+} and negative doubly charged ions of oxygen O^{2-} are formed. Thus, owing to the attraction force arising between the unlike charges, three oxygen ions are bonded with two aluminium ions to form aluminium oxide:

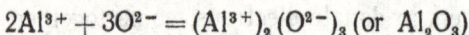

$$Al - 3e^- = Al^{3+}$$
$$O + 2e^- = O^{2-}$$
$$2Al^{3+} + 3O^{2-} = (Al^{3+})_2 (O^{2-})_3 \text{ (or } Al_2O_3)$$

It should be noted that during formation of ionic compounds, *when a great number of positively and negatively charged ions are brought together, the result is a crystal* (*not a molecule*) in which each ion is surrounded by ions of opposite sign, positioned at a distance from it. Obviously, the conception "molecule" in the sense which the gaseous substances would usually imply is inapplicable to the crystal made up of ions (i.e. to ionic compounds in general).

In respect of, for example, sodium chloride we can say that it consists of NaCl molecules only in a very conventional sense. Actually there are no such molecules in its crystals. The entire crystal consists of a large number of Na$^+$ and Cl$^-$ ions. There-fore, the formula NaCl, strictly speaking, does not describe the molecule of sodium chloride. It only indicates that one atom or, more exactly, each chloride ion combines with one atom (ion) of sodium, which defines quite sufficiently the composition of sodium chloride. However, in the future we shall conventionally use the term "molecule" for salts whene-ver we have to indicate the ratio between the numbers of positive and negative ions in them.

The above conceptions of the mechanism of formation of ionic compounds suggest that the *valence of elements in ionic compounds is the number of electrical charges on their ions.* This is otherwise called **electrovalence.**

The value of electrovalence is equal to the number of elect-rons lost by an atom in forming a positive ion or gained by it in forming a negative ion. In the first case the electrovalence is considered positive, and in the second case, negative. For example, in aluminium oxide, aluminium is positively tri-valent, and the oxygen is negatively bivalent. In sodium chlo-ride, sodium is positively monovalent, while chlorine is nega-tively monovalent.

4. Formation of Atomic Compounds

During formation of compounds with atomic (covalent) bonds, the combining atoms neither lose nor gain electrons. The bond between the atoms in a molecule is ensured by the formation of one or several electron pairs which are shared by the neighbouring atoms, that is they belong simultaneously to the electron shells of both atoms and rotate in the orbits embracing the nuclei of both atoms.

Covalent bonds in chemical formulas are denoted in the following manner. Around the symbol of each atom are placed as many points as the atom has valence electrons. The shared electrons are indicated by points placed between the chemical symbols; a double or triple bond is designated respectively by two or three pairs of points placed in between the symbols. Using these designations, we can depict graphically the for-mation and the structure of various molecules with the atomic bonds.

By way of illustration, we give below the schemes of formation of molecules of chlorine and oxygen:

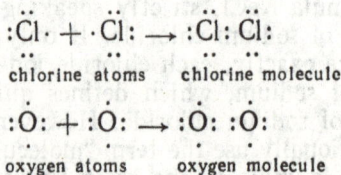

$$:\ddot{Cl}\cdot + \cdot\ddot{Cl}: \longrightarrow :\ddot{Cl}:\ddot{Cl}:$$

chlorine atoms — chlorine molecule

$$:\dot{\ddot{O}}: + :\dot{\ddot{O}}: \longrightarrow :\dot{\ddot{O}}: :\dot{\ddot{O}}:$$

oxygen atoms — oxygen molecule

The following schemes depict the structure of molecules of some chemical compounds with a covalent (atomic) bond:

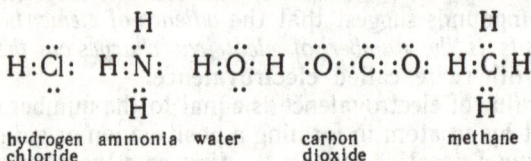

$$H:\ddot{Cl}: \quad H:\ddot{N}: \quad H:\ddot{O}:H \quad :\ddot{O}::C::\ddot{O}: \quad H:\ddot{C}:H$$

hydrogen chloride — ammonia — water — carbon dioxide — methane

Since in the formation of a covalent bond, electrons do not pass from one atom to another, it is obvious that the molecules having the covalent bond have no ions. But if atoms forming a molecule are heterogeneous, the shared electrons can be displaced toward the atom whose nonmetal properties are stronger. For example, in a molecule of HCl, where atoms of hydrogen and chlorine are united by the covalent bond, the shared electrons are shifted toward the chlorine, the element possessing stronger nonmetallic properties, in consequence of which the chlorine atoms acquire a partially negative charge and those of hydrogen a partially positive charge. This can be schematically shown as

$$\overset{+}{H}:\overset{-}{Cl}$$

In such cases the covalent bond is called polar as distinct from the nonpolar covalent bond when the shared electrons are equidistant from both atoms, as is the case with a molecule of hydrogen H : H.

The direction in which the electrons are shifted is determined by the position in the Periodic System of the elements whose atoms compose the given molecule:

(1) inside a period the electrons are shifted from the element which stands to the left toward the element which stands farther to the right; for example in a molecule of PCl_3 the elect-

rons are shifted toward the chlorine which stands to the right
of phosphorus in the same period;

(2) inside the main subgroups the electrons are shifted
from a lower element toward the element which occupies a
higher position; for example, in a molecule of SO_2 the electrons
are shifted toward the oxygen which stands in the same group
as sulphur but above it.

If elements whose atoms form a molecule stand in different periods and groups, the electrons are shifted toward the
element whose nonmetal properties are stronger. For example,
in a molecule of SiF_4 the electrons are shifted toward the fluorine whose nonmetal properties are stronger than in silicon.

5. Valence of Elements in Atomic Compounds

Since in molecules of atomic compounds formed by elements
having different properties some atoms can be charged positively and some negatively (due to the displacement of the shared
electron pairs) the conception of the positive and negative
valence of elements can be extended to the atomic compounds
as well.

*The magnitude of the positive and negative valence of ele-
ments in atomic compounds depends on the charge their ions
would acquire if each of the bonds broke up so that the electrons
forming the bond were transferred completely from one of the
atoms to the other in accordance with the position of the atoms
in the Periodic Table.*

Thus, for example, if in a molecule of PCl_3, the shared
electrons were to pass completely to the atoms of the chlorine,
the latter would have been converted into the negative singly
charged Cl^- ions, and the phosphorus atom, into a positive
triply charged P^{3+} ion. If the shared electrons in a molecule
of SO_2 were transferred to the oxygen atoms, the atoms of the
oxygen would have been converted into negative doubly charged O^{2-} ions, and the sulphur atom, into the positive quarterly
charged S^{4+} ion.

Thus, in the compound PCl_3, the phosphorus is positively
trivalent, and the chlorine is negatively monovalent; in
the compound SO_2 the sulphur is positively tetravalent, while
the oxygen is negatively bivalent.

6. Determination of Valence of Elements in Complex Compounds

From what has been said about the valence of elements in chemical compounds with both ionic and atomic bonds, one can easily arrive at a conclusion that *the algebraic sum of positive and negative valence units of all the constituent atoms in any chemical compound must equal zero.* This enables one to calculate easily the magnitude and the sign of valence of any element in any complex compound, provided the valences of the other constituent elements in this compound are known. One should also bear in mind that hydrogen is positively monovalent in all compounds (except metal hydrides) and oxygen has the valence —2 (except in its compound with fluorine F_2O, where the valence of the oxygen is +2).

The valence of metals is always positive.

Example 1. Determine the valence of manganese in potassium manganate K_2MnO_4.

Solution. From the formula of the compound (K_2MnO_4) it follows that each pair of potassium atoms combines with one manganese and four oxygen atoms. The valence of potassium is +1, and of oxygen —2. By denoting the number of valence units by x, one can make up the equation:

$$2 + x - 2 \times 4 = 0$$

whence

$$x = 8 - 2 = +6$$

that is manganese in this compound has the valence of +6.

Likewise the valence of the elements in various radicals can be determined.

Example 2. Find the valence of phosphorus in the acid radical of pyrophosphoric acid (P_2O_7).

Solution. Since in a molecule of pyrophosphoric acid $H_4P_2O_7$, the radical P_2O_7 is bonded with four atoms of hydrogen whose valence is +1, the valence of the radical is —4 (i.e. the sum of the positive and negative valence units of the atoms forming the radical is —4).

Let the number of valence units in phosphorus be x, then:

$$2x + (-2 \times 7) = -4$$

whence

$$2x = -4 + 14 = +10$$
$$x = +5$$

Hence, the valence of phosphorus is +5.

PROBLEMS

268. How many electrons do the atoms and ions given below donate and accept in transformations indicated by the arrows?

$Cr \longrightarrow Cr^{3+}$, $Fe^{3+} \longrightarrow Fe$, $Sn^{2+} \longrightarrow Sn^{4+}$, $S^{2-} \longrightarrow S$, $Ni^{2+} \longrightarrow Ni^{3+}$

Write the electronic equations for these transformations.

269. How many electrons must Br^-, S^{2-}, Ba^{2+}, Sn^{4+} and Al^{3+} gain or yield in order to be converted into neutral atoms?

270. If two electrons are taken from Te^{2-}, Sn^{2+}, Pt^{4+}, Sb^{3+} and Tl^+ ions what happens to them?

271. Write the electronic equations for the formation of the following ionic compounds from atoms:

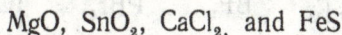

$$MgO, \; SnO_2, \; CaCl_2, \; and \; FeS$$

272. Write the electronic equations for the formation of the following substances:

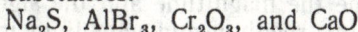

$$Na_2S, \; AlBr_3, \; Cr_2O_3, \; and \; CaO$$

273. What ions are used in the construction of the following crystals:

$$Ag_2O, \qquad MgBr_2, \qquad Fe_2O_3, \qquad and \; CaS$$

Write their symbols.

274. Write out separately all ions contained in CaS, Li_2O, $SnCl_4$, CuO, Bi_2O_3 and AgI.

275. Determine, in the direction of what element the electrons in the molecules of atomic compounds given below must be shifted

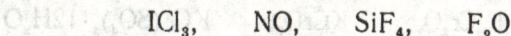

$$ICl_3, \qquad NO, \qquad SiF_4, \qquad F_2O$$

and find the valence of each element.

276. Determine the magnitude and the sign of the valence in compounds SiF_4 and NI_3. Prove your answer.

277. Make out the structure of molecules of the following atomic compounds:

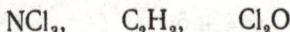

$$NCl_3, \qquad C_2H_2, \qquad Cl_2O$$

Find the valence of the elements in each compound.

278. Make out the structure of molecules:

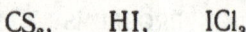

$$CS_2, \qquad HI, \qquad ICl_3$$

Find the magnitude and the sign of the valence for the atoms in each molecule.

279. Draw schemes for the structure of the outer electron shells of the following ions:

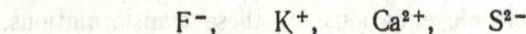

$$F^-, \qquad K^+, \qquad Ca^{2+}, \qquad S^{2-}$$

280. Depict the structure of the following molecules:

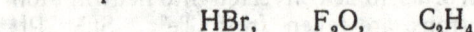

$$HBr, \qquad F_2O, \qquad C_2H_4$$

281. Determine the magnitude and the sign of the valence of elements in the following compounds:

$$SiCl_4, \qquad BF_3, \qquad PBr_3, \qquad ICl$$

282. How many electrons are there in the filled first, third and fourth electron shells of the atom?

283. Determine the magnitude and the sign of the valence of iron, chromium, rhenium and vanadium in the following compounds:

$$K_2FeO_4, \qquad NaCrO_2, \qquad HReO_4, \qquad Ca(VO_3)_2$$

284. Determine the magnitude and the sign of the valence of chlorine, manganese, aluminium and iodine in compounds:

$$HClO_4, \qquad K_2MnO_4, \qquad Zn(AlO_2)_2, \qquad KIO_3$$

285. Find the valence of chromium in the following compounds:

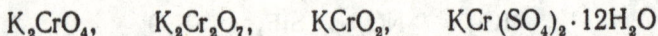

$$K_2CrO_4, \qquad K_2Cr_2O_7, \qquad KCrO_2, \qquad KCr(SO_4)_2 \cdot 12H_2O$$

286. Prove by calculations that the valence of silicon in the naturally occurring compounds like feldspar $K_2Al_2Si_6O_{16}$, kaolin $H_4Al_2Si_2O_9$ and asbestos $CaMg_3Si_4O_{12}$ is the same.

CHAPTER X

CHEMICAL EQUILIBRIUM

1. Rate of Chemical Reactions.
Calculations of Equilibrated Systems

The rate or velocity of a chemical reaction is measured by the change in concentration of the reactants per unit time. By the concentration is understood a quantity of substance per unit volume measured by the number of moles of a substance contained in one litre.

The main factor affecting the velocity of a chemical reaction is the concentration of the reactants and the temperature.

The rate of a chemical reaction is proportional to the product of the concentrations of the reactants. This relationship is known as the **law of mass action,** or the **law of acting masses** and is expressed mathematically as

$$v = k\,[A]^m \cdot [B]^n$$

where v is the velocity of the reaction, $[A]$ and $[B]$ are the concentrations of the reactants A and B, m and n are the coefficients in formulas in the reaction equations, k is a constant for a given reaction called the *velocity constant*.

If the coefficients (m and n) equal 1, the equation is simplified to

$$v = k\,[A] \cdot [B]$$

The law of acting masses holds strictly only for the gaseous substances and solutions. If solid substances also participate in the reaction, their concentration being constant, the velocity of the reaction will only depend on the concentration of the gases or dissolved substances. For example, the rate of the reaction of sulphur burning ($S + O_2 = SO_2$) is proportional to the concentration of oxygen:

$$v = k\,[O_2]$$

The dependence of the reaction rate on the temperature is as follows: *the rate of reaction increases 2-3 times when the temperature is raised* 10° C. As the temperature drops, the rate of a reaction decreases accordingly. The factor showing the number of times the rate of a reaction increases when the temperature is raised 10° is called the *temperature coefficient of the reaction*.

In the course of time the rate of the reaction is reduced since the concentrations of the initial substances decrease in the reaction. But if the reaction is reversible and proceeds in a closed vessel, the rate of the reverse or back reaction (v_2) increases alongside with the decreasing rate of the forward reaction (v_1) which is due to the accumulation of the products of the forward reaction. When both rates are equalized, a state of equilibrium is attained and the concentrations of all reacting substances no longer change.

If we express the reversible or incomplete reaction by the general equation

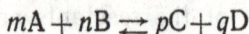

$$mA + nB \rightleftarrows pC + qD$$

the relationship between the concentrations at the state of equilibrium will be expressed by the formula

$$K = \frac{[C]^p \cdot [D]^q}{[A]^m \cdot [B]^n}$$

where K is the constant for a given temperature, called the *equilibrium constant*, which does not depend on the concentration of the reacting substances *. This formula shows that in reversible reactions equilibrium is reached when the product of the concentrations of the resultants divided by the product of the concentrations of the reactants equals a certain constant value for the reaction in question at a given temperature.

The above regularities make it possible to carry out a number of very important calculations.

Example 1. Mixed are 1 litre of glucose and 2 litres of ethyl alcohol. The concentrations of both reactants before mixing

* The constant can also be expressed as $K_1 = \frac{[A]^m[B]^n}{[C]^p[D]^q}$ but in this case the magnitude of the constant will evidently be different: $K_1 = \frac{1}{K}$. In our problems the magnitude of K will always be given for the former case.

were 0.6 mole/litre each. What are the concentrations of the alcohol and glucose after mixing?

Solution. The volume of the mixture is 3 litres. After mixing this volume contains 0.6 mole of glucose and $0.6 \times 2 = 1.2$ moles of ethyl alcohol. Hence, the concentration of glucose is $\frac{0.6}{3} = 0.2$ mole/litre, and that of the alcohol is $\frac{1.2}{3} = 0.4$ mole/litre.

Example 2. How will the rate of the reaction between sulphurous anhydride and oxygen

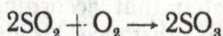

$$2SO_2 + O_2 \longrightarrow 2SO_3$$

change, if the volume of the gaseous mixture is reduced three times?

Solution. Let the concentrations of the sulphurous anhydride and oxygen before the volume has changed be

$$[SO_2] = a \quad \text{and} \quad [O_2] = b$$

In these conditions the rate of the reaction is

$$v = k \cdot a^2 b$$

After the volume has reduced three times, the concentrations of the sulphurous anhydride and oxygen increased accordingly and became:

$$[SO_2] = 3a \quad \text{and} \quad [O_2] = 3b$$

The rate of the reaction between the gases having these concentrations will be

$$v' = k (3a)^2 \cdot 3b = k \cdot 27 a^2 b$$

Comparing v and v' one can see that the rate of the reaction has increased 27 times.

Example 3. The reversible reaction is expressed by the equation:

$$A + 2B \rightleftarrows C$$

At the state of equilibrium the concentrations of the reacting substances are: $[A] = 0.6$ mole/litre, $[B] = 1.2$ moles/litre, $[C] = 2.16$ moles/litre. What is the equilibrium constant and what were the initial concentrations of substances A and B?

Solution. For the given reaction, the equilibrium constant is expressed by the equation

$$K = \frac{[C]}{[A] \cdot [B]^2}$$

By substituting into this equation we obtain

$$K = \frac{2.16}{0.6 \times (1.2)^2} = 2.5$$

To determine the original concentrations of the reactants A and B, it should be noted that according to the equation of the reaction one mole of substance C is formed from one mole of substance A and two moles of substance B. It follows therefore that 2.16 moles of substance A and $2.16 \times 2 = 4.32$ moles of substance B combine to form 2.16 moles of substance C. Thus, the original concentrations of the reactants A and B (that is the number of moles of substances A and B per litre of the mixture before the reaction has begun) were:

$$[A] = 0.6 + 2.16 = 2.76 \text{ moles/litre}$$
$$[B] = 1.2 + 4.32 = 5.52 \text{ moles/litre}$$

Example 4. Mixed are 8 moles of sulphurous anhydride SO_2 and 4 moles of oxygen O_2. The reaction proceeds in a closed vessel at a constant temperature. By the moment the state of equilibrium is attained, the mixture contains 20 per cent of the initial sulphurous anhydride. What is the pressure of the gaseous mixture in equilibrium, if the initial pressure was 3 atm?

Solution. The reaction can be expressed by the equation:

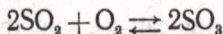

$$2SO_2 + O_2 \rightleftarrows 2SO_3$$

According to the condition of the problem, by the moment when equilibrium was reached, 20 per cent of the initial sulphurous anhydride (1.6 moles) remained in the mixture unreacted. Hence the number of the reacted moles of SO_2 is 6.4. Since according to the equation of the reaction two moles of sulphurous anhydride combine with one mole of oxygen, the number of the reacted moles of oxygen should be 3.2 and that of unreacted, 0.8. The number of moles of sulphuric anhydride formed in the reaction is equal to the number of the reacted moles of the sulphurous anhydride, i.e. 6.4. Thus, the total number of moles of all the three substances at equilibrium will be

$$1.6 + 0.8 + 6.4 = 8.8$$

The pressure of a gas in a closed vessel at constant temperature is directly proportional to the number of moles of the gas present in the vessel. The initial pressure on the gas mixture was 3 atm, while the number of moles was 12. The pressure at equilibrium (p) can be found from the proportion:

$$12 : 8.8 = 3 : p$$

whence

$$p = \frac{8.8 \times 3}{12} = 2.2 \text{ atm or } 222,900 \text{ N/sq m}$$

Example 5. The equilibrium constant for the reaction between carbon dioxide and hydrogen

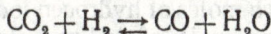

$$CO_2 + H_2 \rightleftharpoons CO + H_2O$$

at a temperature of 850° C is 1. The original concentrations of the reactants were:

$$[CO_2] = 0.2 \text{ mole/litre} \quad \text{and} \quad [H_2] = 0.8 \text{ mole/litre}$$

Calculate the concentrations of all the four substances at the moment when equilibrium is established.

Solution. Let the number of moles of carbon dioxide reacted by the moment of establishing equilibrium be x (per litre of the mixture). From the equation it follows that x moles of hydrogen must also be reacted by this time. As a result, the same number of moles of carbon monoxide and water must be produced in the reaction. Therefore, the concentrations of all the four substances at equilibrium can be expressed as this:

$$[CO] = [H_2O] = x, \quad [CO_2] = 0.2 - x,$$
$$[H_2] = 0.8 - x$$

By substituting these magnitudes into the formula of the equilibrium constant for the given reaction we get

$$\frac{x^2}{(0.2 - x)(0.8 - x)} = 1$$

By solving this equation we find that $x = 0.16$ mole/litre. Hence, at equilibrium

$$[CO] = [H_2O] = 0.16 \text{ mole/litre}$$
$$[CO_2] = 0.2 - 0.16 = 0.04 \text{ mole/litre}$$
$$[H_2] = 0.8 - 0.16 = 0.64 \text{ mole/litre}$$

Example 6. When heated, hydrogen iodide dissociates into iodine and hydrogen. At a certain temperature, the equilibrium constant of this reaction is 1/64. Calculate, how many per cent of hydrogen iodide dissociates at this temperature.

Solution. The dissociation of hydrogen iodide is expressed by the equation:

$$2HI \rightleftarrows H_2 + I_2$$

Assume that the initial concentration of the hydrogen iodide was 2 moles/litre, and by the moment equilibrium was established, x moles out of each two moles of hydrogen iodide dissociated. Since two moles of hydrogen iodide yield one mole of hydrogen and one mole of iodine, x moles of hydrogen iodide must yield $x/2$ moles of hydrogen and the same quantity of iodine. Hence, the concentrations of hydrogen iodide, hydrogen and iodine at equilibrium can be expressed as follows:

$$[HI] = (2 - x) \text{ mole/litre}, \quad [H_2] = [I_2] = \frac{x}{2} \text{ mole/litre}$$

Now substitute these magnitudes into the formula of the equilibrium constant for the given reaction:

$$\frac{1}{64} = \frac{\left(\frac{x}{2}\right)^2}{(2 - x)^2}$$

Solving this equation we obtain:

$$\frac{1}{8} = \frac{x}{2(2 - x)}$$

whence

$$x = 0.4 \text{ mole/litre}$$

Thus, by the moment equilibrium is established, 0.4 mole/litre of hydrogen iodide (out of each two moles) dissociates, which makes 20 per cent of the initial quantity of the hydrogen iodide.

2. Displacement of Chemical Equilibrium

The state of chemical equilibrium depends largely upon the following three factors: (a) concentrations of the reactants, (b) temperature and (c) pressure (if gases or vapours participate in the reaction). If any of the three factors changes, the equilibrium is upset and the concentrations of all the reacting substances start changing as well. This change in the concentrations continues until the rates of the forward and back reactions are equalized in the new conditions. Then equilibrium is restored, but now with different concentrations of all substances.

The change in concentrations caused by the disturbance of equilibrium is called a *displacement* or *shift* of equilibrium. If the concentrations of the substances in the right half of the equation increase, the equilibrium is said to shift to the right, if the concentrations undergo the opposite change, we say that the equilibrium has shifted to the left.

The direction in which equilibrium is shifted can be determined by using the following rule:

A change in the conditions of an equilibrated system, such as temperature, pressure or concentration, will shift the equilibrium in the direction of the reaction opposing change (Le Chatelier 's principle).

For example, in the reaction

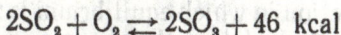

$$2SO_2 + O_2 \rightleftarrows 2SO_3 + 46 \text{ kcal}$$

the equilibrium will be shifted as follows:

1. With growing concentration of the sulphurous anhydride or oxygen, and also with reducing concentration of the sulphuric anhydride, the equilibrium will shift to the right since in this condition the concentrations of the sulphurous anhydride and oxygen will reduce again, whereas the concentration of the sulphuric anhydride will grow. And vice versa, the reduced concentration of the sulphurous anhydride or oxygen, or else the increased concentration of the sulphuric anhydride, will shift the equilibrium to the left to increase the concentrations of the sulphurous anhydride and oxygen alongside with reducing concentration of the sulphuric anhydride.

2. With lowering temperature, the equilibrium is shifted to the right, since the reaction of formation of sulphuric anhyd-

ride is accompanied by the evolution of heat, and hence the said shift in the equilibrium will raise the temperature again. Conversely, if the temperature is raised, the equilibrium will be shifted to the left in the direction of the endothermal reaction.

3. If the pressure is increased by compressing the reaction mixture, the equilibrium will be shifted to the right, since with this shift, the pressure will reduce again due to the reduction of the total number of molecules in the mixture (two molecules of sulphurous anhydride and one molecule of oxygen yield only two molecules of sulphuric anhydride). The reduced pressure will shift the equilibrium to the left which will be accompanied by the increase in the total number of molecules, owing to which the pressure will grow again.

In all these cases, the immediate cause of the upset equilibrium is the disturbance of the equality of the rates of the forward and back reactions.

If the rates of the forward and back reactions change equally with the variations in the temperature, concentration or pressure, the equilibrium will not be upset. But if the rate of the forward reaction increases or diminishes greater than the rate of the back reaction, the equilibrium will be upset and shifted in the direction of the reaction whose rate became greater. This shift will continue until the rates of both reactions are equalized.

Thus, the direction in which equilibrium is shifted can often be determined by calculating the changes in the rates of the forward and back reactions.

Example. On heating a mixture of carbon monoxide with chlorine in a closed vessel the following equilibrium is established

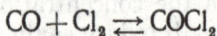

$$CO + Cl_2 \rightleftarrows COCl_2$$

How will the rates of the forward and the back reactions change if (at constant temperature) the pressure is increased two times by reducing the volume of the gas mixture? Will this change shift the equilibrium?

Solution. Let the concentrations of carbon monoxide, chlorine and phosgene at equilibrium be

$$[CO] = a, \qquad [Cl_2] = b, \qquad [COCl_2] = c$$

In these conditions the rate of the forward reaction is

$$v_f = k \cdot ab$$

while the rate of the back reaction is

$$v_b = k_1 c$$

After the volume of the gaseous mixture has been reduced two times, the concentrations of all gases also increased two times and the rates of the forward and back reactions became

$$v_f' = k \cdot 2a \cdot 2b = k \cdot 4ab$$
$$v_b' = k_1 \cdot 2c$$

By dividing v_f' by v_f we can find how much the rate of the forward reaction has become greater

$$\frac{v_f'}{v_f} = \frac{k \cdot 4ab}{k \cdot ab} = 4$$

Likewise, the increase in the rate of the back reaction can be found:

$$\frac{v_b'}{v_b} = \frac{k_1 \cdot 2c}{k_1 \cdot c} = 2$$

Thus, the rate of the forward reaction increases four times, and that of the back reaction only two times. Since equilibrium can be attained only when the rates of both reactions are equal, it is evident that in the example under question, the equilibrium will be upset and displaced toward the reaction whose rate is now greater, that is in the direction of formation of phosgene.

It is easy to make clear that if Le Chatelier's principle is used in the solution of this problem the result will be the same.

PROBLEMS

287. The concentration of a gas is 3 moles/litre. What is the pressure on the gas if its temperature is 0° C?

288. Express in mole/litre the concentration of a gas at STP.

289. Assuming that atmospheric air contains 20 per cent of oxygen (by volume), what is the concentration (in mole/litre) of oxygen in atmospheric air at STP?

290. Mixed are 2 litres of substance A and 3 litres of substance B. The concentration of substance A before mixing was 0.6 mole/litre, and that of substance B 1 mole/litre. Calculate the concentrations of both substances at the first moment after mixing.

291. Write mathematical expressions for the rates of these reactions:

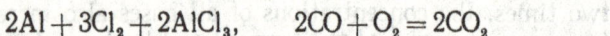

$$2Al + 3Cl_2 + 2AlCl_3, \qquad 2CO + O_2 = 2CO_2$$

292. A gaseous mixture consists of hydrogen and chlorine. The reaction proceeds according to the equation:

$$H_2 + Cl_2 = 2HCl$$

How will the rate of the reaction change if the pressure is tripled?

293. The interaction between carbon monoxide and chlorine proceeds according to the equation

$$CO + Cl_2 \rightleftarrows COCl_2$$

The concentration of carbon monoxide is 0.3 mole/litre, and that of chlorine 0.2 mole/litre. How will the rate of the forward reaction change if the concentration of the chlorine increases to 0.6 mole/litre and that of the carbon monoxide to 1.2 moles/litre?

294. How will the rate of the reaction

$$2NO + O_2 \rightarrow 2NO_2$$

proceeding in a closed vessel change if the pressure is increased four times?

295. The reaction between substances A and B is represented by the equation

$$A + 2B = C$$

The initial concentration of substance A is 0.3 mole/litre and that of substance B 0.5 mole/litre. The rate constant of the reaction is 0.4. What is the initial rate of the reaction and that in a lapse of a certain time when the concentration of substance A reduces by 0.1 mole?

296. One gram-molecule of gas A and two gram-molecules of gas B were introduced into one vessel and two gram-molecules of gas A and one gram-molecule of gas B were placed into another vessel. The temperature in both vessels is the same. Will the rate of the reaction between gases A and B in both vessels be equal if the reaction is expressed by the equations

$$\text{(a)} \quad A + B = C$$
$$\text{(b)} \quad 2A + B = D$$

297. At a temperature of $150°$ C a certain reaction is completed in 16 minutes. Assuming the temperature coefficient of the reaction to be 2.5, what time in minutes is required to complete the same reaction at a temperature of $200°$ C and $80°$ C?

298. The equilibrium of the reaction $H_2 + I_2 \rightleftarrows 2HI$ has been established at the following concentrations of the reacting substances:

$$[H_2] = 0.25 \text{ mole/litre}, \qquad [I_2] = 0.05 \text{ mole/litre},$$
$$[HI] = 0.9 \text{ mole/litre}$$

What were the initial concentrations of the iodine and hydrogen?

299. At equilibrium in the system

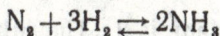

$$N_2 + 3H_2 \rightleftarrows 2NH_3$$

the concentrations of the reacting substances are:

$$[N_2] = 3 \text{ moles/litre}, \qquad [H_2] = 9 \text{ moles/litre},$$
$$[NH_3] = 4 \text{ moles/litre}$$

What were the initial concentrations of the hydrogen and nitrogen?

300. On heating nitrogen dioxide in a closed vessel to a certain temperature, equilibrium of the reaction

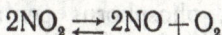

$$2NO_2 \rightleftarrows 2NO + O_2$$

was established at the following concentrations of the reacting substances:

$$[NO_2] = 0.06 \text{ mole/litre}, \qquad [NO] = 0.24 \text{ mole/litre},$$
$$[O_2] = 0.12 \text{ mole/litre}$$

Find the equilibrium constant for the given temperature and the initial concentration of nitrogen dioxide.

301. The starting substances in the reaction

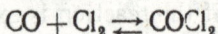

$$CO + Cl_2 \rightleftarrows COCl_2$$

are taken in equivalent quantities. The reaction proceeds in a closed vessel at a constant temperature. By the moment of establishing equilibrium, 50 per cent of the initial quantity of carbon monoxide remain unreacted. What is the pressure at equilibrium if the initial pressure of the mixture was 1 atm?

302. How will the pressure change by the moment of establishing equilibrium in the reaction

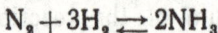

$$N_2 + 3H_2 \rightleftarrows 2NH_3$$

proceeding in a closed vessel at a constant temperature, if the initial concentrations of the nitrogen and hydrogen are 2 and 6 moles/litre respectively and if the equilibrium sets when 10 per cent of the initial nitrogen has reacted?

303. The reversible reaction is represented by the equation:

$$A + B \rightleftarrows C + D$$

The equilibrium constant is 1. The initial concentration of substance A is 2 moles/litre. Calculate how many per cent of substance A react if the initial concentrations of substance B are 2, 10 and 20 moles/litre.

304. The reversible reaction is expressed by the equation:

$$A + B \rightleftarrows C + D$$

The equilibrium constant is 2.5. Can equilibrium be attained if the concentrations of all four substances are equal? Justify your answer.

305. When heated, phosphorus pentachloride dissociates according to the equation

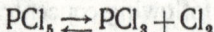

$$PCl_5 \rightleftarrows PCl_3 + Cl_2$$

At a certain temperature 1.5 moles out of 2 moles of PCl_5 in a closed 10-litre vessel are decomposed. Calculate the equilibrium constant at this temperature.

306. If a mixture of carbon dioxide and hydrogen is heated in a closed vessel, equilibrium is established:

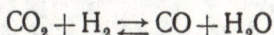

$$CO_2 + H_2 \rightleftarrows CO + H_2O$$

The equilibrium constant at 850° C is 1. What quantity in per cent of carbon dioxide converts into carbon monoxide at 850° C if one gram-molecule of carbon dioxide is mixed with five gram-molecules of hydrogen?

307. The equilibrium constant for the reaction

$$H_2 + I_2 \rightleftarrows 2HI$$

at 445° C is 50. How many moles of hydrogen per one mole of iodine are required to convert 90 per cent of the iodine into hydrogen iodide?

308. On heating a mixture of carbon dioxide and hydrogen to 850° C, equilibrium was established

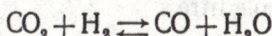

$$CO_2 + H_2 \rightleftarrows CO + H_2O$$

The equilibrium constant at this temperature is 1. What was the volume ratio of the carbon dioxide to hydrogen before mixing if by the moment of establishing equilibrium 90 per cent of hydrogen were converted into water?

309. At 1000° C the equilibrium constant of the reaction

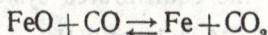

$$FeO + CO \rightleftarrows Fe + CO_2$$

is 0.5. What are the equilibrium concentrations of the carbon monoxide and carbon dioxide, if their initial concentrations were:

$$[CO] = 0.05 \text{ mole/litre} \quad \text{and} \quad [CO_2] = 0.01 \text{ mole/litre}$$

310. In certain conditions hydrogen chloride is oxidized by oxygen to form water vapour and chlorine at the following equilibrium:

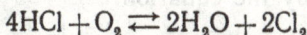

$$4HCl + O_2 \rightleftarrows 2H_2O + 2Cl_2$$

The reaction is exothermic. Answer the following questions:

(a) In what direction will the equilibrium shift with growing temperature?

(b) Will the equilibrium be upset if the pressure is increased at constant temperature?

311. Why in the reaction $N_2 + 3H_2 \rightleftarrows 2NH_3$ the equilibrium will, and in the reaction $N_2 + O_2 = 2NO$ will not be upset with changing pressure?

312. The reduction of carbon dioxide with carbon is represented by the equation

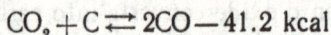

$$CO_2 + C \rightleftarrows 2CO - 41.2 \text{ kcal}$$

In what direction will the equilibrium be shifted with growing temperature? Will it be upset with growing pressure?

313. Equilibrium of the reaction

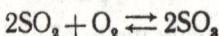

$$2SO_2 + O_2 \rightleftarrows 2SO_3$$

has been set at the following concentrations of the reacting substances:

$[SO_2] = 0.1$ mole/litre, $[O_2] = 0.05$ mole/litre,
$[SO_3] = 0.9$ mole/litre

Calculate, how the rate of the forward and the back reactions will change if the volume occupied by the gases is reduced two times. Will the equilibrium be upset? Prove your answer by the appropriate calculations.

314. Calculate, how the rates of the forward and the back reactions change in the equilibrated systems:

$$H_2 + I_2 \rightleftarrows 2HI$$

and

$$2NO + O_2 \rightleftarrows 2NO_2$$

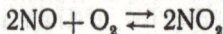

if the pressure is doubled.

Basing on the results of the calculations, decide how the said change in the pressure will affect the equilibrium in these systems.

315. The formation of ammonia from hydrogen and nitrogen is expressed by the equation:

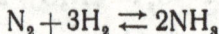

$$N_2 + 3H_2 \rightleftarrows 2NH_3$$

Basing on the equality of the rates of the forward and the back reactions at equilibrium, prove that with increasing pressure the equilibrium will be displaced in the direction of formation of ammonia.

316. In what direction will the equilibria

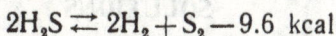

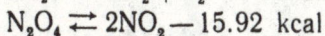

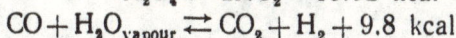

$$2H_2S \rightleftarrows 2H_2 + S_2 - 9.6 \text{ kcal}$$
$$N_2O_4 \rightleftarrows 2NO_2 - 15.92 \text{ kcal}$$
$$CO + H_2O_{vapour} \rightleftarrows CO_2 + H_2 + 9.8 \text{ kcal}$$

be displaced (a) at lowering temperature? (b) at increasing pressure?

317. How will elevated pressure at constant temperature affect the equilibrium in the following systems?

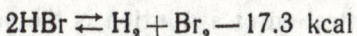

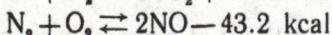

$$2HBr \rightleftarrows H_2 + Br_2 - 17.3 \text{ kcal}$$
$$2CO + O_2 \rightleftarrows 2CO_2 + 135.2 \text{ kcal}$$
$$N_2 + O_2 \rightleftarrows 2NO - 43.2 \text{ kcal}$$

In what direction will these equilibria be displaced with increasing temperature?

318. How will growing pressure affect the following equilibria?

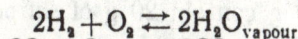

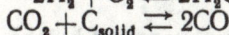

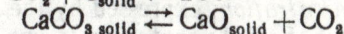

$$2H_2 + O_2 \rightleftarrows 2H_2O_{vapour}$$
$$CO_2 + C_{solid} \rightleftarrows 2CO$$
$$CaCO_{3\ solid} \rightleftarrows CaO_{solid} + CO_2$$

319. What changes in the concentrations of the reacting substances can shift the equilibrium of the reaction

$$CO_2 + C \rightleftarrows 2CO$$

to the right?

CHAPTER XI

SOLUTIONS

1. Concentration of Solutions

The concentration of a solution is the quantity by weight of solute contained in a definite weight or in a definite volume of the solution.

There are three methods of expressing concentration in the current chemical practice. These are:

(a) *Percent concentration*, expressed in per cent by weight, indicating the number of grams of solute contained in 100 g of solution. For example, a 20 per cent solution of saltpetre is a solution whose 100 g contain 20 g of saltpetre and 80 g of water.

(b) *Molar concentration*, expressed by the number of gram-molecules (moles) of solute contained in one litre of solution. Solutions with their concentrations expressed in this manner are called molar. They are denoted by the letter M preceded by a coefficient indicating the "molarity" of the solution, that is the number of moles per litre of solution. For example, $1M$ solution contains one mole of solute in one litre, a $0.1M$ solution contains 0.1 mole of solute per litre of solution, etc.

(c) *Normality*, expressed by the number of gram-equivalents of solute contained in one litre of solution. A solution that contains one gram-equivalent of solute per litre is called a uninormal or simply normal solution and is denoted as $1N$. If the solution contains 0.5 gram-equivalent per litre it is called seminormal ($0.5N$), etc.

In determining the quantity of a substance required to prepare a solution of the wanted normality, the following should be borne in mind.

The *equivalent weight of an acid* equals its molecular weight divided by its basicity. For example the equivalent weight of phosphoric acid H_3PO_4 is

$$\frac{H_3PO_4}{3} = \frac{98}{3} = 32.7$$

The *equivalent weight of a base* equals its molecular weight divided by the valence of the metal in it. For example, the equivalent weight of calcium hydroxide $Ca(OH)_2$ is

$$\frac{Ca(OH)_2}{2} = \frac{74}{2} = 37$$

The *equivalent weight of a salt* equals its molecular weight divided by the valence of the metal and by the number of its atoms in a molecule of the salt. For example, the equivalent weight of $Fe_2(SO_4)_3$ is

$$\frac{Fe_2(SO_4)_3}{3 \times 2} = \frac{400}{6} = 66.7$$

The concentration of a solution is sometimes characterized by its density. Since the density of a solution depends on the concentration of solute, by determining the density of a solution one can find its concentration from the appropriate tables.

A. Preparation of Solutions of Various Concentrations

Calculations connected with the preparation of solutions of various concentrations can be illustrated by the following examples.

Example 1. How many grams of sodium hydroxide are required to prepare three litres of a 10 per cent solution?

Solution. By consulting the corresponding table we find the density of a 10 per cent solution of sodium hydroxide. It is 1.109 g/ml.

The mass of three litres of the solution can be found from the relationship between mass m, volume V and density ρ

$$m = V \cdot \rho = 3{,}000 \times 1.109 = 3{,}327 \text{ g}$$

According to the condition of the problem, 10 per cent of this mass belongs to the sodium hydroxide. Hence, the quantity of sodium hydroxide required to prepare three litres of a 10 per cent solution is

$$\frac{3{,}327 \times 10}{100} = 332.7 \text{ g}$$

Example 2. What weight in grams of blue vitriol $CuSO_4 \cdot 5H_2O$ is required to prepare one kilogram of an 8 per cent solution calculated as the anhydrous salt?

Solution. One kilogram of an 8 per cent solution should contain $1,000 \times 0.08 = 80$ g of the anhydrous salt. One gram-molecule of $CuSO_4$ weighs 160 g, and one gram-molecule of blue vitriol $CuSO_4 \cdot 5H_2O$ weighs 250 g. It follows therefore that 250 g of blue vitriol contain 160 g of the anhydrous salt.

The quantity of blue vitriol containing 80 g of the anhydrous salt can be found from the proportion

$$250 : 160 = x : 80$$

Solving for x:

$$x = \frac{250 \times 80}{160} = 125 \text{ g}$$

Thus, to prepare one kilogram of an 8 per cent solution, 125 g of blue vitriol and $1,000 - 125 = 875$ g of water are required.

Example 3. What weight in grams of nitric acid is contained in 200 ml of a $0.1M$ solution?

Solution. One gram-molecule of nitric acid weighs 63 g. Hence, one litre of a $0.1M$ solution contains 6.3 g of HNO_3 and 200 ml of the solution contain

$$\frac{6.3 \times 200}{1,000} = 1.26 \text{ g}$$

Example 4. What is the molarity (i.e. the number of moles per litre) of a solution whose 300 ml contain 10.5 g of potassium hydroxide?

Solution. Let us find the quantity of potassium hydroxide contained in one litre of the solution from the proportion

$$300 : 1000 = 10.5 : x$$
$$x = 35 \text{ g}$$

A gram-molecule of potassium hydroxide weighs 56 g. It follows therefore that 35 g make $\frac{35}{56} = 0.625$ gram-molecule.

The molarity of the solution is 0.625.

Example 5. What weight in grams of soda $Na_2CO_3 \cdot 10H_2O$ is required to prepare one litre of a $0.1N$ solution?

Solution. The molecular weight of soda $Na_2CO_3 \cdot 10H_2O$ is 286, whence one gram-equivalent of soda is $286 : 2 = 143$ g and 0.1 gram-equivalent is 14.3 g. It follows therefore that to prepare one litre of a $0.1N$ solution, 14.3 g of crystalline soda should be dissolved in a small quantity of water in a 1-litre measuring flask and then water should be added to the mark.

Example 6. How many millilitres of water should be added to 200 ml of a 68 per cent solution of nitric acid (density, 1.4 g/ml) in order to obtain a 10 per cent solution?

Solution. The mass of 200 ml of a 68 per cent solution of nitric acid is $1.4 \times 200 = 280$ g. It contains $280 \times 0.68 = 190.4$ g of HNO_3. The same quantity of the acid will be contained in a dilute solution, making only 10 per cent of its mass. Hence, the mass of the entire solution after adding water will be

$$\frac{190.4 \times 100}{10} = 1,904 \text{ g}$$

Since the mass of the initial solution is 280 g, the quantity of water which should be added to obtain the required concentration of the solution is

$$1,904 - 280 = 1,624 \text{ g}$$

Example 7. How many millilitres of a 96 per cent sulphuric acid (density 1.84 g/ml) should be taken to prepare one litre of a $0.5N$ solution?

Solution. The equivalent weight of sulphuric acid is 1/2 its molecular weight, i.e. 49. To prepare one litre of a $0.5N$ solution, 0.5 gram-equivalent or 24.5 g of sulphuric acid are required.

The mass of one millilitre of a 96 per cent sulphuric acid is 1.84 g. It contains $\frac{1.84 \times 96}{100} = 1.77$ g of H_2SO_4. Hence, to prepare one litre of a $0.5N$ solution, the required quantity of the 96 per cent sulphuric acid is

$$\frac{24.5}{1.77} = 13.84 \text{ ml}$$

Example 8. There is a $2N$ solution of barium hydroxide $Ba(OH)_2$. How can one litre of a $0.1N$ solution be prepared out of it?

Solution. One litre of a $2N$ solution of barium hydroxide contains 2 gram-equivalents of $Ba(OH)_2$, and one litre of a $0.1N$ solution contains 0.1 gram-equivalent of $Ba(OH)_2$. This quantity of $Ba(OH)_2$ will be contained in $\frac{1,000 \times 0.1}{2} =$ $=50$ ml of the $2N$ solution. It follows therefore that to prepare one litre of a $0.1N$ solution of barium hydroxide 50 ml of the $2N$ solution should be diluted with water to make one litre.

B. Relationship Between the Three Units of Concentration

During chemical calculations it is often necessary to convert the percent concentration into the molar or normal concentration and vice versa. The examples which follow exemplify such recalculations.

Example 1. What is the molarity of a 20 per cent solution of hydrochloric acid having the density of 1.10 g/ml?

Solution. The mass of one litre of a 20 per cent solution of hydrochloric acid HCl is 1,100 g. It contains $1,100 \times 0.20 =$ $= 220$ g of HCl. A gram-molecule of HCl weighs 36.5 g. To determine the molarity of a 20 per cent solution we first find how many moles of HCl compose 220 g:

$$220 : 36.5 = 6.03 \text{ moles}$$

The molarity of the solution is 6.03.

Example 2. What is the percent concentration of a $2N$ solution of sodium hydroxide having the density of 1.08 g/ml?

Solution. The molecular weight of sodium hydroxide is 40. One litre of a $2N$ solution has the mass of $1.08 \times 1,000 =$ $= 1,080$ g and contains two moles or 80 g of NaOH. The percent concentration of sodium hydroxide in the solution can be found from the proportion:

$$1,080 : 80 = 100 : x$$

Solving for x:

$$x = 7.4 \text{ per cent}$$

C. Calculation of Reacting Volumes

When the concentrations of the solutions used in various reactions are known, one can easily calculate the required volumes of the reactants that should be mixed to ensure complete reaction of the substances. These calculations are even more simplified if the solutions have definite normalities. Since the weights of the reactants are proportional to their equivalent weights, it is evident that solutions containing equal numbers of gram-equivalents of solute should always be taken for the reaction. If normalities of such solutions are equal, their volumes will also be the same. If otherwise, the volumes will relate in the reverse proportion to the normalities.

Let the volumes of solutions spent in the reaction be V_1 and V_2 and their normalities, that is the concentrations expressed in gram-equivalents, be C_1 and C_2 respectively. Then:

$$V_1 : V_2 = C_2 : C_1$$

or

$$V_1 \cdot C_1 = V_2 \cdot C_2$$

Example 1. How many millilitres of a $0.25N$ solution of sulphuric acid are required to precipitate all barium (as $BaSO_4$) from 20 ml of a $2N$ solution of barium chloride?

Solution. Let the sought volume of the sulphuric acid be V. The proportion appears then as this:

$$V : 20 = 2 : 0.25$$

whence

$$V = \frac{20 \times 2}{0.25} = 160 \text{ ml}$$

Example 2. In order to neutralize 42 ml of an acid, 14 ml of a $0.3N$ solution of an alkali were added. What is the normality of the acid?

Solution. The problem can be solved according to the model of the previous example. Let the normality of the acid be x. Then:

$$x : 0.3 = 14 : 42$$

Solving for x:

$$x = \frac{0.3 \times 14}{42} = 0.1$$

PROBLEMS

320. Express in per cent the concentration of a solution containing 40 g of sugar in 280 g of water.

321. What weight in grams of borax $Na_2B_4O_7 \cdot 10H_2O$ and how much water are required to prepare two kilograms of a 5 per cent solution of $Na_2B_4O_7$ (calculated with reference to the anhydrous salt)?

322. How many grams of sodium sulphite Na_2SO_3 are required to prepare five litres of an 8 per cent solution having the density of 1.075 g/ml?

323. What weight in grams of sodium nitrate $NaNO_3$ is required to prepare 300 ml of a 0.2M solution?

324. What weight in grams of Glauber salt $Na_2SO_4 \cdot 10H_2O$ is required to prepare two litres of a 0.5N solution of sodium sulphate?

325. How many grams of soda Na_2CO_3 are contained in 500 ml of a 0.25N solution?

326. How many millilitres of water should be added to 100 ml of a 20 per cent solution of H_2SO_4 (density 1.14 g/ml) to obtain a 5 per cent solution of the acid?

327. What will be the percent concentration of nitric acid if to 500 ml of a 32 per cent acid having the density of 1.2 g/ml one litre of water is added?

328. To what volume should 500 ml of a 20 per cent solution of sodium chloride (density 1.152 g/ml) be diluted to obtain a 4.5 per cent solution having the density of 1.029 g/ml?

329. How many millilitres of a 2N solution of sulphuric acid are required to prepare 500 ml of a 0.5N solution?

330. How many millilitres of water should be added to 100 ml of a 1N solution in order to prepare a 0.05N solution?

331. There is a 2M solution of soda Na_2CO_3. How can one litre of a 0.25N solution be prepared out of it?

332. How many millilitres of concentrated hydrochloric acid containing 38 per cent of HCl (density 1.19 g/ml) are required to prepare one litre of a 2N solution?

333. To 100 ml of a 96 per cent sulphuric acid (density 1.84 g/ml) 400 ml of water were added. A solution having the density of 1.225 g/ml was obtained as a result. Express its concentration in per cent and in gram-equivalents per litre.

334. Calculate the normality of concentrated hydrochloric acid (density 1.18 g/ml) containing 36.5 per cent of HCl.

335. How many millilitres of a 10 per cent sulphuric acid (density 1.07 g/ml) are required to neutralize a solution containing 16 g of NaOH?

336. There are a solution whose one litre contains 18.9 g of HNO_3 and a solution containing 3.2 g of NaOH per one litre. What volumes of these solutions should be mixed to obtain a neutral solution?

337. How many millilitres of a $0.2N$ solution of an alkali are required to precipitate all iron (as $Fe(OH)_3$) from 100 ml of a $0.5N$ solution of $FeCl_3$?

338. What weight in grams of calcium carbonate will be precipitated from 400 ml of a $0.5N$ solution of calcium chloride if excess soda solution is added?

339. To neutralize 20 ml of a $0.1M$ solution of an acid, 8 ml of sodium hydroxide solution were spent. How many grams of the sodium hydroxide are contained in one litre of this solution?

340. To neutralize 30 ml of a $0.1N$ solution of an alkali 12 ml of an acid solution were spent. What is the normality of the acid?

341. To neutralize 40 ml of an alkali solution, 24 ml of a $0.5N$ solution of sulphuric acid were spent. What is the normality of the alkali solution? How many millilitres of a $0.5N$ solution of hydrochloric acid will be required to obtain the same result?

342. Calculate the equivalent weights of lime $Ca(OH)_2$ and phosphoric acid H_3PO_4. Calculate from the equivalent weights how many grams of lime are required to neutralize 150 g of phosphoric acid. Calculate also from the equation of the reaction.

343. Calculate the equivalent weights of sulphuric acid and aluminium hydroxide. Find from the equivalent weights, how many millilitres of sulphuric acid are required to convert 65 g of aluminium hydroxide into aluminium sulphate. Calculate it also from the equation of the reaction.

344. In order to neutralize a solution containing 2.25 g of an acid, 25 ml of a $2N$ alkali solution were spent. Determine the equivalent weight of the acid.

345. To neutralize 20 ml of a solution containing 12 g of an alkali per litre, 24 ml of a $0.25N$ solution of an acid were spent. Calculate the equivalent weight of the alkali.

2. Solubility

The measure of solubility of a substance is the concentration of its saturated solution. This concentration is very often expressed by the number of grams of a solute contained in 100 g of a solvent.

Solubility of most solids increases with temperature, whereas solubility of gases would always decrease with growing temperature. The dependence of solubility on temperature is usually expressed graphically by solubility curves, the temperature being plotted against the x-axis and solubility against the y-axis. By using the solubility curves one can easily determine solubility of various substances at a given temperature.

In addition to temperature, the solubility of gases also depends on the pressure.

The mass of a gas dissolved in a given volume of a liquid is directly proportional to the pressure of the gas. But since with growing or decreasing pressure, the mass of a unit volume of a gas increases (decreases) accordingly, one can come to a conclusion that *the volume of gas dissolvable in a given volume does not depend on the pressure.* For this reason, the solubility of gases is usually expressed by the number of volumes of a gas that can be dissolved at a given temperature in 100 volumes of a solvent.

If there is a mixture of several gases above the liquid, the solubility of each gas depends on its partial pressure.

Below are given some examples of calculations which require the knowledge of solubility of substances.

Example 1. To purify potassium nitrate KNO_3 by crystallization, 300 g of the salt were dissolved with heating in 200 g of water. How many grams of the salt will crystallize out from the solution cooled to 10°C?

Solution. From the solubility curves shown in Fig. 2 one can find that the solubility of the salt at 10°C is about 22 g per 100 g of water or 44 g per 200 g of water. It follows there-

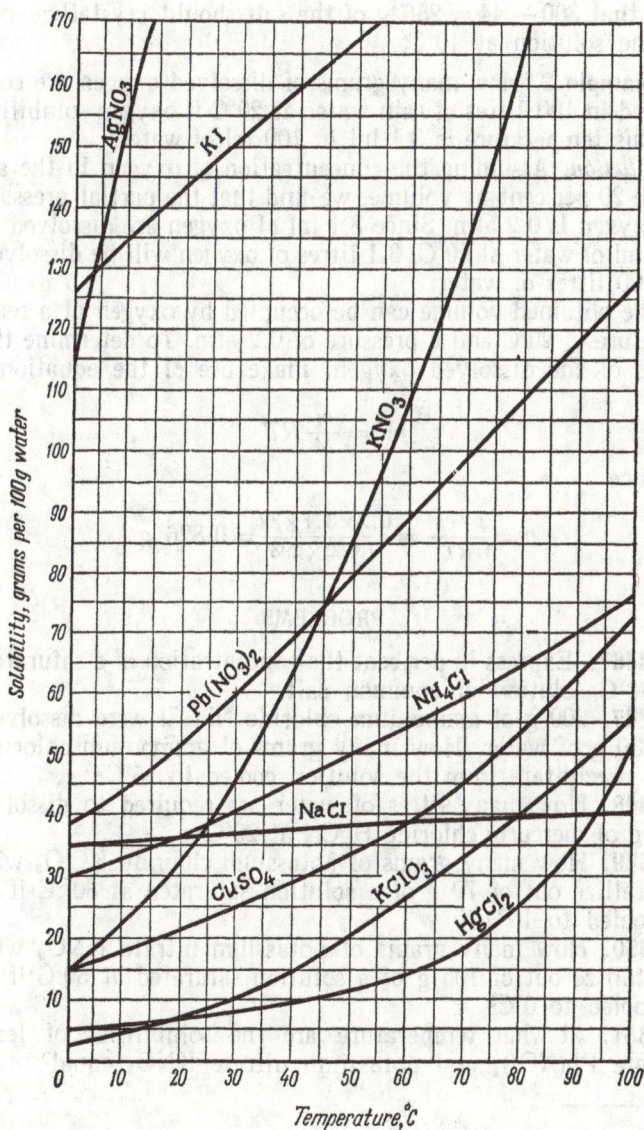

Fig. 2. Solubility curves of various salts

fore that $300 - 44 = 256$ g of the salt should crystallize out of the solution at 10°C.

Example 2. How many grams of dissolved oxygen are contained in 100 litres of rain water at 20°C if oxygen solubility at this temperature is 3.1 ml in 100 ml of water?

Solution. Assuming the concentration of oxygen in the air to be 20 per cent by volume, we find that the partial pressure of oxygen is 0.2 atm. Since 3.1 ml of oxygen are dissolved in 100 ml of water at 20°C, 3.1 litres of oxygen will be dissolved in 100 litres of water.

The obtained volume can be occupied by oxygen at a temperature of 20°C and a pressure of 0.2 atm. To determine the mass of the dissolved oxygen, make use of the equation

$$pV = \frac{m}{M} RT$$

whence

$$m = \frac{pV \cdot M}{RT} = \frac{0.2 \times 3.1 \times 32}{0.082 \times 293} = 0.826 \text{ g}$$

PROBLEMS

346. Express in per cent the concentration of a saturated at 20°C solution of common salt.

*347. 300 g of ammonium chloride NH_4Cl were dissolved in 500 g of water. How many grams of ammonium chloride will precipitate from the solution cooled to 15°C?

*348. How many litres of water are required to dissolve 100 g of mercuric chloride $HgCl_2$ at 20°C?

*349. How many grams of potassium chlorate $KClO_3$ will crystallize out of 70 g of a solution saturated at 80° C if it is cooled to 10° C?

*350. How many grams of potassium nitrate KNO_3 will crystallize out of 105 g of a solution saturated at 60°C if it is cooled to 0°C?

*351. At what temperature are the solubilities of lead nitrate $Pb(NO_3)_2$ and potassium nitrate KNO_3 equal?

---

* In solving problems marked with an asterisk use should be made of the solubility curves given in Fig. 2.

*352. At what temperature will a 20 per cent solution of copper sulphate $CuSO_4$ be saturated?

353. There is a saturated at 20°C solution of blue vitriol. Will it remain so when cooled to a temperature below 20°C, provided crystals fall out of the solution? Will it remain saturated when heated to a temperature above 20°C?

354. A litre of water is saturated with carbon dioxide at 0°C and a pressure of 5 atm. What volume will the dissolved carbon dioxide occupy if it is separated from the water and reduced to standard conditions of pressure and temperature? (The solubility of carbon dioxide at 0°C is 171 ml in 100 ml of water.)

355. Solubility of ammonia at 20°C is 702 ml per 1 ml of water. Express the concentration of the resulting solution in per cent.

356. 4.62 litres of hydrogen sulphide are dissolved in one litre of water at 0°C. At what pressure should the hydrogen sulphide be dissolved in water to prepare its 5 per cent solution?

357. Assuming that the air contains 21 per cent by volume of oxygen and 79 per cent of nitrogen, what is the percent composition of the air liberated from water having the temperature of 20°C? The solubility of oxygen at this temperature is 31 ml and that of nitrogen 15.4 ml per litre of water.

358. A gaseous mixture containing 40 per cent of nitrous oxide and 60 per cent of nitric oxide (by volume) was dissolved at a temperature of 17°C and constant pressure in water to complete saturation. What will be the percent composition of the mixture after its separation from the water, if solubility of nitrous oxide at 17°C is 690 ml and that of nitric oxide is 50 ml per litre of water?

3. Heats of Solution and Hydration

Dissolution of various substances in water is accompanied with liberation or absorption of heat. During solution of a solid substance its crystal lattice is broken and its entities are distributed through the bulk of the solvent. This process requires energy and is therefore accompanied by the absorption of heat. However, alongside with the dissolution another process, namely, formation of hydrates, often takes place, which is accompanied by the evolution of heat. Depending on which thermal effect is greater, the overall thermal effect will be either positive or negative.

Dissolution of liquids is in most cases associated with liberation of heat. The same thermal effect is observed with the dissolution of gases.

The amount of heat liberated or absorbed when one mole of a substance is dissolved is called the *heat of solution* of a substance. By experimentally determining the heats of solution of an anhydrous solid substance and of its crystal hydrate (if any is possible with the given substance) one can calculate the heat of hydration *.

Example 1. During solution of 10 g of ammonium chloride NH_4Cl in 243 g of water, the temperature lowered 3°C. Determine the heat of solution of ammonium chloride.

Solution. Assuming the specific heat (c) of the solution to be 1 (due to the small concentration), we can calculate from the reduction of the temperature ($\Delta T = -3°C$) the amount of absorbed heat (Q):

$$Q = c \cdot \Delta t = 243(-3) = -729 \text{ cal}$$

A gram-molecule of ammonium chloride weighs 53.5 g. The amount of the absorbed heat is proportional to the quantity of dissolved NH_4Cl. By denoting the heat of solution by q, we can make out the proportion

$$q : -729 = 53.5 : 10$$

whence

$$q = -3,900 \text{ calories} \quad \text{or} \quad -3.9 \text{ kcal/mole}$$
$$\text{or} - 16.33 \text{ kJ/mole}$$

Example 2. During solution of one mole of anhydrous calcium chloride $CaCl_2$, 18.0 kcal are evolved ($q_1 = 18.0$ kcal/mole), and during solution of one mole of the crystal hydrate $CaCl_2 \cdot 6H_2O$; 4.56 kcal are absorbed ($q_2 = -4.56$ kcal/mole). What is the heat of hydration of calcium chloride?

Solution. The amount of heat evolving during dissolution of anhydrous calcium chloride is the algebraic sum of two

---

* Solution heats can vary slightly depending on the quantity of the solvent used.

thermal effects, viz., that of the solution proper and of hydration. Since during solution of the crystal hydrate $CaCl_2 \cdot 6H_2O$ no hydration takes place (the salt has already been hydrated), the magnitude of 18 kcal/mole is the heat of solution of calcium chloride. By denoting the heat of hydration by x we can write

$$q_1 = q_2 + x, \qquad 18 = -4.56 + x$$

whence

$$x = 22.56 \text{ kcal/mole} \quad \text{or} \quad 94.45 \text{ kJ/mole}$$

PROBLEMS

359. On dissolution of 10 g of sodium hydroxide in 250 g of water, the temperature increased 9.71°C. What is the heat of solution of NaOH if the specific heat of the solution is 1?

360. On dissolution of one mole of H_2SO_4 in 800 g of water, the temperature increased 22.4°C. What is the heat of solution of sulphuric acid if the specific heat of the solution is $0.9 \dfrac{cal}{g \cdot deg}$?

361. The heat of solution of ammonium nitrate NH_4NO_3 is -6.4 kcal. By what number of degrees will the temperature decrease when 20 g of ammonium nitrate are dissolved in 180 g of water? The specific heat of the obtained solution is $0.9 \dfrac{cal}{g \cdot deg}$.

362. On dissolution of 32 g of anhydrous copper sulphate $CuSO_4$ in 80 g of water, 3.16 kcal are liberated and during solution of 50 g of its crystal hydrate $CuSO_4 \cdot 5H_2O$ in the same quantity of water, 0.56 kcal is absorbed. What is the heat of hydration of copper sulphate?

363. The heat of solution of crystal hydrate of sodium sulphate $Na_2SO_4 \cdot 10H_2O$ is -18.8 kcal/mole. Calculate the lowering of the temperature on solution of 0.5 mole of this salt in 1,000 g of water assuming that the specific heat of the solution is 1.

364. During solution of 8 g of anhydrous copper sulphate $CuSO_4$ in 192 g of water the temperature increases 3.95°C. What is the heat of hydration of copper sulphate if the heat of solution of its crystal hydrate $CuSO_4 \cdot 5H_2O$ is -2.8 kcal/mole and the specific heat of the solution is 1?

365. Calculate, by what number of degrees the temperature increases during solution of one mole of sulphuric acid in 200 **g** of water, assuming the specific heat of the solution to be 0.75 cal/g·deg and the heat of solution of sulphuric acid 18.1 kcal/mole.

4. Osmotic Pressure of Solution

Each solution is characterized by a definite osmotic pressure whose magnitude in dilute solutions is proportional to the concentration of the solute and the absolute temperature.

The osmotic pressure of a solution equals the pressure which would be exerted by the solute if it were in a gaseous state at the same temperature and occupied a volume equal to that of the solution (the **law of Van't Hoff**).

The dependence of the osmotic pressure on the volume of the solution, the quantity of the solute and the temperature is expressed by the equation of Van't Hoff which is quite similar to the equation of state of a gas:

$$PV = RT$$

where P is the osmotic pressure, V is the volume of the solution containing one gram-molecule of the solute, R is the constant whose numerical value is the same as that of a gas constant, and T is the absolute temperature.

By substituting in this equation $\frac{1}{C}$, (where C is the concentration of the solution expressed in moles per litre), for the gram-molecular volume V we have the following expression convenient for calculating osmotic pressure:

$$P = CRT$$

This equation shows that with constant volume and temperature, the magnitude of osmotic pressure depends only on the number of molecules of the solute, and does not depend on the nature of the solute or the solvent. Therefore, equimolecular solutions (that is solutions containing equal numbers of molecules per litre) of various substances have equal osmotic pressures at the same temperature. In particular, solutions containing one gram-molecule of the solute in 22.4 litres of the solution have an osmotic pressure equal to 1 atm at 0°C.

This rule only does not hold for electrolytes (acids, alkalis and salts), in which the magnitude of osmotic pressure is always greater than theoretical.

The above regularities make it possible to calculate osmotic pressures of nonelectrolyte solutions from their concentration, temperature and the molecular weight of the solute. If osmotic pressure of a solution is known from experiment, then knowing the concentration of the solution, one can calculate the molecular weight of the solute. During calculations it should be borne in mind that the law of osmotic pressure holds strictly only for very dilute solutions, whereas for concentrated solutions the calculations will be only approximate.

Example 1. What is the osmotic pressure of a solution whose one litre contains 0.2 mole of a nonelectrolyte (a) at 0°C? (b) at 17°C?

Solution. The osmotic pressure of a solution containing 1 mole of a solute in one litre at 0°C is theoretically 22.4 atm. Since the osmotic pressure is proportional to the concentration of the solution, the sought osmotic pressure can be found from the following proportion:

$$22.4 : x = 1 : 0.2$$

Solving for x:

$$x = 4.48 \text{ atm}$$

To find the osmotic pressure at 17°C let us make use of the Van't Hoff equation:

$$P = CRT$$

According to the conditions of the problem $C = 0.2$ mole, $T = 290°K$. By substituting into the equation and assuming R to be 0.082 we have:

$$P = 0.2 \times 0.082 \times 290 = 4.76 \text{ atm*}$$

Example 2. Determine the molecular weight of mannite knowing that the osmotic pressure of the solution containing 9 g of mannite in 250 ml of the solution is 4.5 atm at 0°C.

Solution. According to the conditions of the problem, one litre of the solution contains 36 g of mannite or $36/M$ moles,

---

* Or 482,300 N/sq m in the units of the International System.

where M is the sought molecular weight of mannite. By making use of the Van't Hoff equation and knowing that $C = \frac{36}{M}$, we have

$$4.5 = \frac{36}{M} 0.082 \times 273$$

whence

$$M = \frac{36}{4.5} 0.082 \times 273 = 179.1$$

The problem can also be solved by using the proportion:

$$36 : M = 4.5 : 22.4$$

$$M = \frac{36 \times 22.4}{4.5} = 179.1$$

PROBLEMS

366. There are a solution containing 5 g of naphthalene $C_{10}H_6$ in one litre and a solution containing 5 g of anthracene $C_{14}H_{10}$, also in one litre. What solution has greater osmotic pressure? Prove your answer without calculating the osmotic pressures.

367. How do the weights of formaldehyde HCHO and glucose $C_6H_{12}O_6$ relate to each other if they are contained in equal volumes of solutions having equal osmotic pressure at a given temperature?

368. Mixed are one volume of glucose $C_6H_{12}O_6$ solution and three volumes of urea $CO(NH_2)_2$ solution. The osmotic pressure of the former is 2.8 atm and of the latter, 1.4 atm. The temperature of both solutions is the same. What is the osmotic pressure of the mixture at this temperature?

369. There are water solutions of glycerol $C_3H_5(OH)_3$ and ethyl alcohol C_2H_5OH containing equal weights of the solute in one litre. The temperature of the solutions is the same. The osmotic pressure of what solution is greater? How much greater?

370. What is the osmotic pressure of a solution containing 0.05 mole of a solute (nonelectrolyte) in one litre at 35°C?

371. Calculate the osmotic pressure of a solution containing 3.1 g of aniline $C_6H_5NH_2$ in one litre. The temperature of the solution is 21°C.

372. Is the osmotic pressure equal in solutions of nonelectrolytes containing at the same temperature and in equal volumes (a) equal number of grams and (b) equal number of gram-molecules of various solutes?

373. The osmotic pressure of an aqueous solution containing 1 g of sugar $C_{12}H_{22}O_{11}$ in 100 ml is 0.655 atm at 0°C. Prove that for solutions the constant R in the equation of Van't Hoff $PV=RT$ has the same numerical value as for gases.

374. How many gram-molecules of a nonelectrolyte must one litre of a solution contain if its osmotic pressure at 0°C has to be 1 atm?

375. How many grams of glucose $C_6H_{12}O_6$ must one litre of a solution contain if its osmotic pressure has to be the same as of a solution containing 3 g of formaldehyde HCHO in one litre at the same temperature?

376. A solution containing 3.75 g of formaldehyde per litre has the osmotic pressure of 2.8 atm at 0°C. What is the molecular weight of formaldehyde?

377. The osmotic pressure of a solution containing 3 g of sugar in 250 ml of solution is 0.82 atm at 12°C. What is the molecular weight of sugar?

5. Vapour Pressure of Saturated Solutions

On dissolution of a non-volatile substance in a liquid, the pressure of saturated vapour of this liquid decreases. Thus, the vapour pressure of a solution is always lower than the vapour pressure of the pure solvent at the same temperature, the pressure reduction being proportional to the concentration of the solution.

The relationship between lowering of the vapour pressure and the concentration of the solution can be expressed for dilute solutions by the following equation:

$$\Delta p = p \frac{n}{N}$$

where Δp is the lowering of the vapour pressure, that is the difference between the vapour pressure of a pure solvent and of the solution; p is the vapour pressure of a pure solvent; n is the number of moles of the solute and N is the number of moles of the solvent.

The above relationship does not hold for electrolyte solutions. Considerable deviations from this rule are also observed with concentrated solutions.

By making use of the above equation one can (a) calculate vapour pressure of various solutions from the concentration of the solution and the vapour pressure of the pure solvent; (b) calculate molecular weights of solutes if the vapour pressures of the solution and of the pure solvent, and also the concentration of the solution and the molecular weight of the solvent are known; (c) make other calculations connected with lowering of the vapour pressure.

Example 1. What is the vapour pressure of a solution containing 0.2 mole of sugar in 450 g of water at 20°C?

Solution. First it is necessary to calculate how many moles of water are contained in 450 g. Since one mole of water weighs 18 g, the sought number of moles is

$$N = \frac{450}{18} = 25$$

The vapour pressure of pure water (p) at 20°C is 17.5 mm Hg; $n=0.2$ mole.

By substituting into the equation we can find the value of lowering of the vapour pressure:

$$\Delta p = \frac{17.5 \times 0.2}{25} = 0.14 \text{ mm Hg}$$

Hence, the vapour pressure is

17.5 — 0.14 = 17.36 mm Hg, or 2,314 N/sq m

Example 2. At 50°C the vapour pressure of a solution containing 23 g of the solute in 200 g of ethyl alcohol C_2H_5OH is 207.2 mm Hg. The vapour pressure of pure alcohol at the same temperature is 219.8 mm Hg. What is the molecular weight of the solute?

Solution. The lowering of the vapour pressure of alcohol is

$$\Delta p = 219.8 - 207.2 = 12.6 \text{ mm Hg}$$

The molecular weight of alcohol is 46, and the number of

moles of the solvent is

$$N = \frac{200}{46} = 4.35$$

By substituting these values in the equation $\Delta p = p\frac{n}{N}$ we can find the number of moles (n) of the solute:

$$12.6 = \frac{219.8 \cdot n}{4.35}$$

$$n = \frac{12.6 \times 4.35}{219.8} = 0.25 \text{ mole}$$

According to the conditions of the problem, 0.25 mole of the solute weighs 23 g, hence the molecular weight of the solute is

$$M = \frac{23}{0.25} = 92$$

PROBLEMS

378. What is the vapour pressure at 65° C of a solution containing 13.68 g of sugar $C_{12}H_{22}O_{11}$ in 90 g of water, if the vapour pressure of water at the same temperature is 187.5 mm Hg?

379. What is the vapour pressure of a 10 per cent solution of urea $CO(NH_2)_2$ at 100°C?

380. At 42°C the vapour pressure of water is 61.5 mm Hg. How much will the vapour pressure lower at this temperature if 36 g of glucose $C_6H_{12}O_6$ are dissolved in 540 g of water?

381. At 20°C the vapour pressure of water is 17.5 mm Hg. How many grams of glycerol $C_3H_5(OH)_3$ should be dissolved in 180 g of water to lower the vapour pressure 1 mm Hg?

382. The vapour pressure of water at 70°C is 233.8 mm Hg. At the same temperature the vapour pressure of a solution containing 12 g of a solute in 270 g of water is 230.68 mm Hg. What is the molecular weight of the solute?

383. Determine the molecular weight of aniline knowing that at 30°C the vapour pressure of a solution containing 3.09 g of aniline in 370 g of ether $(C_2H_5)_2O$ is 643.6 mm Hg and the vapour pressure of the pure ether at the same temperature is 647.9 mm Hg.

6. Freezing and Boiling of Solutions

All pure liquids are characterized by strictly definite freezing and boiling points. The presence of a solute raises the boiling point and lowers the freezing point of the solvent. Therefore, solutions freeze at lower and boil at higher temperatures than pure solvents.

The following principles hold true for dilute nonelectrolyte solutions with respect to their freezing and boiling points:

1. *The lowering of the freezing point and raising of the boiling point are proportional to the quantity of substance dissolved in a given mass of the solvent.*

2. *Equimolecular* (that is containing equal numbers of moles) *quantities of various substances dissolved in the same mass of a given solvent lower its freezing point and raise its boiling point an equal number of degrees* (laws of Raoult).

The lowering of the freezing point (by calculation) due to dissolution of one mole of a substance in 1,000 g of a solvent is a constant for a given solvent. It is called the *cryoscopic constant* of the solvent. By analogy, the elevation of a boiling point caused by dissolving 1 gram-molecule of a substance in 1,000 g of a solvent is called the *ebullioscopic constant* of the solvent. The cryoscopic and ebullioscopic constants are different for various solvents.

Cryoscopic and Ebullioscopic Constants of Some Solvents

Solvent	$E_{f.\,p.}$, °C	$E_{b.\,p.}$, °C
Water	1.86	0.52
Benzene	5.2	2.57
Ethyl alcohol	—	1.16
Diethyl ether	—	2.12

The relationship between the magnitude of the lowering of the freezing point of a solvent and the amount of a substance dissolved in a solution can be expressed mathematically as this:

$$\Delta t_{\text{f. p.}} = E_{\text{f. p.}} \cdot C$$

where $\Delta t_{\text{f.p.}}$ is the lowering of the freezing point of a solvent

$E_{\text{f.p.}}$ is the cryoscopic constant of the solvent

C is the number of moles of a solute in 1,000 g of a solvent.

Since the number of moles is equal to the mass of the substance in grams (m) divided by its molecular weight (M), by substituting m/M for C we obtain:

$$\Delta t_{\text{f. p.}} = E_{\text{f. p.}} \cdot \frac{m}{M}$$

A similar formula expresses the relationship between elevation of the boiling point of a solvent and the amount of substance dissolved in it:

$$\Delta t_{\text{b. p.}} = E_{\text{b. p.}} \cdot \frac{m}{M}$$

where $E_{\text{b.p.}}$ is the ebullioscopic constant of the solvent.

The above formulas can be used to calculate the freezing and boiling points of solutions of nonelectrolytes from their concentrations, and also to determine molecular weights of solutes from the freezing and boiling points of their solutions. Like in the case with the osmotic pressure, the calculations for concentrated solutions are only approximate.

Neither can these formulas be employed for solutions of electrolytes unless appropriate corrections are known.

Example 1. What is the approximate temperature at which a solution containing 54 g of glucose $C_6H_{12}O_6$ in 250 g of water will freeze if the cryoscopic constant of water is 1.86°C?

Solution. The amount of glucose in 1,000 g of water will be 216 g. First let us calculate the number of moles contained in 216 g of glucose. Since its molecular weight is known to be 180, the sought number of moles is

$$C = 216 : 180 = 1.2$$

Now calculate the lowering of the freezing point:

$$\Delta t_{\text{f. p.}} = E_{\text{f. p.}} \cdot C = 1.86 \times 1.2 = 2.23°\text{C}$$

Hence, the solution will freeze at —2.23°C.

Example 2. When 0.94 g of phenol was dissolved in 50 g of ethyl alcohol, the boiling point of the alcohol elevated 0.232°C. What is the molecular weight of phenol if the ebullioscopic constant of alcohol is 1.16°C?

Solution. As in the previous example, we first calculate the number of grams of phenol contained in 1,000 g of the alcohol:

$$0.94:50 = x:1,000$$
$$x = 18.8 \text{ g}$$

If one gram-molecule of phenol had been dissolved in 1,000 g of alcohol, the boiling point would have risen 1.16°C. Since the elevation of the boiling point (with the same amount of the solvent) is proportional to the mass of the dissolved substance, by designating the sought molecular weight of phenol as M, we can derive the proportion:

$$0.232:1.16 = 18.8:M$$

whence

$$M = 94 \text{ g}$$

that is the molecular weight of phenol is 94.

Example 3. A solution containing 8 g of naphthalene $C_{10}H_8$ in 100 g of diethyl ether boils at 36.33°C while pure ether boils at 35°C. What is the ebullioscopic constant of the ether?

Solution. From the conditions of the problem we find that

$$\Delta t_{b.\,p.} = 36.33 - 35 = 1.33° C$$
$$M = 128; \quad C = 80:128 = 0.625 \text{ mole}$$

By substituting into the formula

$$\Delta t_{b.\,p.} = E_{b.\,p.} \cdot C$$

we have

$$1.33 = E_{b.\,p.} \cdot 0.625$$

whence

$$E_{b.\,p.} = \frac{1.33}{0.625} = 2.128° C$$

PROBLEMS

384. Solutions of different nonelectrolytes containing equal amounts of solute in equal volumes will freeze at the same temperature, in what conditions?

385. There are a 10 per cent solution of methyl alcohol CH_3OH and a 10 per cent solution of ethyl alcohol C_2H_5OH. Which of them will freeze at a lower temperature?

386. How many degrees will the boiling point elevate if 9 g of glucose $C_6H_{12}O_6$ are dissolved in 100 g of water?

387. At what (approximately) temperature will a 50 per cent solution of sugar $C_{12}H_{22}O_{11}$ boil?

388. At what (approximately) temperature will a 40 per cent solution of ethyl alcohol C_2H_5OH freeze?

389. How many degrees will the freezing point of benzene C_6H_6 lower, if 4 g of naphthalene $C_{10}H_8$ are dissolved in its 100 g?

390. How many grams of sugar $C_{12}H_{22}O_{11}$ should be dissolved in 100 g of water in order (a) to lower the freezing point of water 1°C? (b) to raise its boiling point 1°C?

391. In what ratio by weight should water and ethyl alcohol C_2H_5OH be mixed to prepare a mixture freezing at —20°C?

392. Into an automobile radiator were primed 9 litres of water and 2 litres of methyl alcohol CH_3OH (density 0.8 g/ml). At what lowermost temperature can now an automobile be parked in the open?

393. When 5 g of a substance are dissolved in 200 g of water, a dielectric solution freezing at —1.55°C is obtained. What is the molecular weight of the substance?

394. A solution containing 6 g of urea in 50 g of water freezes at a temperature of —3.72°C. What is the molecular weight of urea?

395. When 15 g of chloroform were dissolved in 400 g of diethyl ether, the boiling point was raised 0.665°C. What is the molecular weight of chloroform?

396. When 13 g of camphor were dissolved in 400 g of diethyl ether, the boiling point was raised 0.453°C. What is the molecular weight of camphor?

397. When 8.9 g of anthracene $C_{14}H_{10}$ were dissolved in 200 g of ethyl alcohol, the boiling point was raised 0.29°C. Determine the ebullioscopic constant of the alcohol.

398. A solution containing 2.7 g of phenol C_6H_5OH in 75 g of benzene C_6H_6 freezes at a temperature of 3.5°C, while pure benzene freezes at 5.5°C. What is the cryoscopic constant of benzene?

399. When 6.18 g of naphthalene were dissolved in 150 g of benzene the freezing point lowered 1.67°C. What is the molecular weight of naphthalene?

400. When 3.24 g of sulphur were dissolved in 40 g of benzene the boiling point of the solvent elevated 0.81°C. Of how many atoms does a molecule of sulphur consist in the solution?

401. What is the formula of a substance containing 50.69 per cent of carbon, 4.23 per cent of hydrogen and 45.08 per cent of oxygen if a solution containing 2.13 g of this substance in 60 g of benzene freezes at 4.25°C? The freezing point of pure benzene is 5.5°C.

CHAPTER XII

SOLUTIONS OF ELECTROLYTES

1. Electrolytic Dissociation
(Ionization)

When dissolved in water, acids, bases and salts fall into positively and negatively charged particles, the ions. This process is called electrolytic dissociation or ionization. It is reversible, and therefore an equilibrium is established between ions and non-dissociated molecules.

Molecules of acids break up into positively charged hydrogen ions and negatively charged ions of acid radicals. Molecules of bases disintegrate into positively charged metal ions and negatively charged hydroxyl ions, and finally salts break up into metal ions and acid radical ions.

The ratio of the number of ionized molecules to the total number of molecules dissolved is called the *degree of ionization or dissociation*, and designated by the Greek letter α. For example, if 80 out of 200 molecules of the dissolved substance have fallen into ions, the degree of dissociation is $80 : 200 = 0.4$ or 40 per cent. The degree of dissociation increases with dilution of the solution with water.

Depending on the degree of dissociation, the electrolytes would be usually divided into *strong* and *weak*. Strong electrolytes are those in which the degree of ionization in dilute solutions is great and its variations with growing concentration are comparatively insignificant. These are the majority of salts, alkalis (except NH_4OH), and some acids (HCl, HNO_3, H_2SO_4 and others). In weak electrolytes the degree of ionization is, on the contrary, very low (even in very dilute solutions) and it decreases rapidly with growing concentration.

Polybasic acids dissociate in a somewhat peculiar way: their molecules dissociate *stepwise*, first one hydrogen ion being split, then the second, etc. For example, dissociation of phosphoric acid H_3PO_4 consists of the following steps:

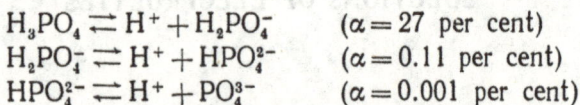

$$H_3PO_4 \rightleftarrows H^+ + H_2PO_4^- \qquad (\alpha = 27 \text{ per cent})$$
$$H_2PO_4^- \rightleftarrows H^+ + HPO_4^{2-} \qquad (\alpha = 0.11 \text{ per cent})$$
$$HPO_4^{2-} \rightleftarrows H^+ + PO_4^{3-} \qquad (\alpha = 0.001 \text{ per cent})$$

As follows from the above data, which refer to a 0.1 N solution of H_3PO_4, the dissociation is stronger in the first equation (first dissociation), in the second equation it is weaker and is only insignificant in the third equation.

Acid salts dissociate in steps as well. For example:

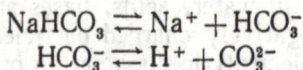

$$NaHCO_3 \rightleftarrows Na^+ + HCO_3^-$$
$$HCO_3^- \rightleftarrows H^+ + CO_3^{2-}$$

However, the degree of ionization in the second step is very small, therefore the solution of the acid salt only contains an insignificant quantity of hydrogen ions.

Basic salts dissociate into ions of the basic and acid radicals. For example:

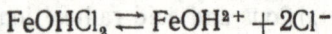

$$FeOHCl_2 \rightleftarrows FeOH^{2+} + 2Cl^-$$

The second dissociation, that is dissociation of ions of basic radicals into metal and hydroxyl ions, almost does not take place.

The degree of ionization of an electrolyte in a given solution can be determined by various methods, in particular from the osmotic pressure of the solution, lowering of the vapour pressure, lowering of the freezing point or raising of the boiling point. All these values are known to depend on the number of particles of the solute contained in a given quantity of the solvent. Since due to the dissociation, the number of entities of the solute (molecules and ions) in a solution of an electrolyte is greater than in an equimolar solution of a nonelectrolyte, the osmotic pressure and all other related values will evidently be greater in a solution of an electrolyte.

A factor showing the number of times the experimentally found values (osmotic pressure, lowering of vapour pressure, and others) exceed the values calculated theoretically from the molar concentration of the same solution is called the *isotonic coefficient*. The isotonic coefficient i is expressed by the following ratios:

$$i = \frac{P'}{P} = \frac{\Delta p'}{\Delta p} = \frac{\Delta t'_{f.\,p.}}{\Delta t_{f.\,p.}} = \frac{\Delta t'_{b.\,p.}}{\Delta t_{b.\,p.}}$$

where P' is the osmotic pressure; $\Delta p'$ is the lowering of vapour pressure; $\Delta t'_{f\,p.}$ is the lowering of freezing point and $\Delta t'_{b.p.}$ is the elevation of the boiling point observed actually with the solution of the electrolyte; P, Δp, $\Delta t_{f.p.}$ and $\Delta t_{b.p.}$ are theoretically calculated magnitudes of the same values. Since all these values are proportional to the number of entities of the solute, *the isotonic coefficient equals the ratio of the number of entities in the solution (ions and undissociated molecules) to the total number of molecules of the solute.*

There is a relationship between the isotonic coefficient i, the degree of ionization α and the number of ions n, into which a molecule of an electrolyte breaks up, which is expressed by the formula:

$$\alpha = \frac{i-1}{n-1}$$

It is obvious from the formula that to determine the degree of ionization, one must only find the magnitude of the coefficient i for a given solution.

Making use of the above formula one can calculate the degree of ionization for various electrolytes from the osmotic pressure, from the lowering of the vapour pressure or from the elevation of the boiling point or lowering of the freezing point of the solutions. And vice versa, knowing the degree of ionization of a dissolved electrolyte and the concentration of the solution, one can easily determine any of the above values.

According to the modern views, strong electrolytes in aqueous solutions completely dissociate into ions, but in experimental determination (by the electric conductivity of a solution or by some other methods) the degree of ionization is always less than 100 per cent. This is explained by the electrostatic interaction between ions, which reduces the activity of ions, and all properties of solutions which depend on the concentration of ions show themselves as if the number of ions in the solution were less than that with complete dissociation of the electrolyte.

As a solution of a strong electrolyte is diluted with water, it is not the degree of ionization but the activity of the ions that is intensified due to the increased mean distance between them. Therefore, when measured by conventional methods, the degree of ionization of strong electrolytes is only "apparent", and not true. Nevertheless it is an important characteristic of an electrolyte which determines many of its properties.

Example 1. A solution of 0.85 g of zinc chloride $ZnCl_2$ in 125 g of water freezes at $-0.23°C$. What is the apparent degree of ionization of zinc chloride in this solution?

Solution. We first express the concentration of the solution in moles per 1,000 g of water. Since the molecular weight of zinc chloride is 136, it follows that

$$C = \frac{0.85 \times 1,000}{125 \times 136} = 0.05 \text{ mole/1,000 g } H_2O$$

Now we find the theoretical magnitude of the lowering of the freezing point $\Delta t_{f.p.}$ for the given solution from the formula

$$\Delta t_{f.\,p.} = E_{f.\,p.} \cdot C$$

Since for water $E_{f.p.} = 1.86$, we have

$$\Delta t_{f.\,p.} = 1.86 \times 0.05 = 0.093°C$$

Now we can calculate the isotonic coefficient i, bearing in mind that from the conditions of the problem $\Delta t'_{f.p.} = 0.23°C$:

$$i = \frac{\Delta t'_{f.\,p.}}{\Delta t_{f.\,p.}} = \frac{0.23}{0.093} = 2.47$$

Then the apparent degree of ionization is

$$\alpha = \frac{i-1}{n-1} = \frac{2.47-1}{2} = 0.735 \text{ or } 73.5 \text{ per cent}$$

Example 2. The apparent degree of ionization of zinc sulphate $ZnSO_4$ in a $0.1N$ solution is 40 per cent. What is the osmotic pressure of the solution at 0°C?

Solution. We first calculate the theoretical magnitude of the osmotic pressure from the equation

$$P = CRT$$

Since the equivalent weight of zinc sulphate is half its molecular weight, a $0.1N$ solution contains 0.05 mole of $ZnSO_4$

per litre. Hence, $C = 0.05$. By substituting into the equation the numerical values for C, R and T, we get

$$P = 0.05 \times 0.082 \times 273 = 1.12 \text{ atm}$$

To calculate the actual osmotic pressure P', determine the coefficient i for the given solution. To this end use the formula for calculation of the degree of ionization.

In accordance with the conditions of the problem $\alpha = 0.4$. A molecule of zinc sulphate dissociates into two ions, hence $n = 2$. It follows then

$$0.4 = \frac{i-1}{2-1} \qquad i = 1.4$$

Now, since $i = \dfrac{P'}{P}$,

$$P' = P \cdot i = 1.12 \times 1.4 = 1.57 \text{ atm} \quad \text{or} \quad 159,080 \text{ N/sq m}$$

Example 3. What is the vapour pressure at 100°C of a solution containing 5 g of sodium hydroxide in 180 g of water, if the apparent degree of ionization of sodium hydroxide in this solution is 80 per cent?

Solution. We first find the coefficient i for the given solution. In accordance with the conditions of the problem $\alpha = 0.8$ and $n = 2$. Hence

$$0.8 = \frac{i-1}{2-1} \qquad i = 1.8$$

Now calculate the theoretical lowering of the vapour pressure of the solution Δp from the equation

$$\Delta p = p \frac{n}{N}$$

At 100°C the vapour pressure of water is 760 mm Hg. The molecular weight of sodium hydroxide is 40, and that of water is 18. Hence, $n = 5 : 40 = 0.125$; $N = 180 : 18 = 10$. By substituting into the equation we get

$$\Delta p = \frac{760 \times 0.125}{10} = 9.5 \text{ mm Hg}$$

The actual lowering of the vapour pressure $\Delta p'$ can be found from the formula

$$i = \frac{\Delta p'}{\Delta p}$$

$$1.8 = \frac{\Delta p'}{9.5}, \text{ whence } \Delta p' = 17.1 \text{ mm Hg}$$

The sought vapour pressure of the solution

$$p' = p - \Delta p' = 760 - 17.1 = 742.9 \text{ mm Hg or } 99{,}040 \text{ N/sq m}$$

Example 4. What is the isotonic coefficient of a 0.2 M solution of an electrolyte if its one litre contains 2.19×10^{23} entities (molecules and ions) of the dissolved substance?

Solution. Since the total number of molecules and ions of the electrolyte is 2.19×10^{23}, and the number of molecules taken for the preparation of the 0.2 M solution is evidently $6.02 \times 10^{23} \times 0.2 = 1.20 \times 10^{23}$, it follows that by dividing the first magnitude by the second one, we shall have the isotonic coefficient

$$i = \frac{2.19 \times 10^{23}}{1.20 \times 10^{23}} = 1.82$$

PROBLEMS

402. Write the equations for dissociation into ions of the following substances:

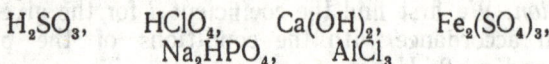

$$H_2SO_3, \qquad HClO_4, \qquad Ca(OH)_2, \qquad Fe_2(SO_4)_3,$$
$$Na_2HPO_4, \qquad AlCl_3$$

403. Into what ions do the following substances dissociate:

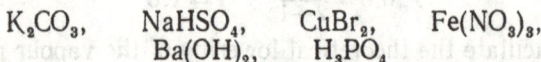

$$K_2CO_3, \qquad NaHSO_4, \qquad CuBr_2, \qquad Fe(NO_3)_3,$$
$$Ba(OH)_2, \qquad H_3PO_4$$

404. Write the equations for ionization of these substances:

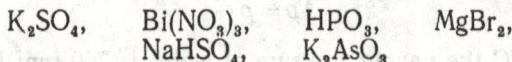

$$K_2SO_4, \qquad Bi(NO_3)_3, \qquad HPO_3, \qquad MgBr_2,$$
$$NaHSO_4, \qquad K_2AsO_3$$

405. On dissolution of 0.01 mole of acetic acid CH_3COOH in water, 20 per cent of molecules have broken into ions. How many separate entities of the dissolved substance does the solution contain?

406. Are the osmotic pressures of molar solutions of glucose $C_6H_{12}O_6$, potassium nitrate KNO_3, ethyl alcohol C_2H_5OH and acetic acid CH_3COOH equal? Justify your answer.

407. A molar solution of hydrocyanic acid HCN freezes almost at the same temperature as a molar solution of sugar $C_{12}H_{22}O_{11}$. What conclusion can be made with respect to the degree of ionization of hydrocyanic acid?

408. There are $0.1M$ solutions of aluminium chloride $AlCl_3$ and calcium chloride $CaCl_2$. The apparent degree of ionization of both salts in these solutions is approximately the same. What solution will freeze at a lower temperature? Why?

409. There are decimolar solutions of acetic acid CH_3COOH, sodium chloride $NaCl$, glucose $C_6H_{12}O_6$ and calcium chloride $CaCl_2$. Arrange them in order of decreasing osmotic pressure.

410. There are solutions of calcium nitrate $Ca(NO_3)_2$ and ferric sulphate $Fe_2(SO_4)_3$ containing equal number of moles per litre of water. What solution has greater osmotic pressure, if the apparent degree of ionization of the dissolved salts is the same?

411. A solution containing 2.1 g of potassium hydroxide in 250 g of water freezes at $-0.519°C$. What is the isotonic coefficient of the solution?

412. Determine the isotonic coefficient for a solution of magnesium chloride containing 0.1 mole of $MgCl_2$ in 1,000 g of water if the solution freezes at $0.461°C$.

413. When a gram-molecule of potassium nitrate KNO_3 was dissolved in one litre of water the freezing point lowered $3.01°C$. Determine the apparent degree of ionization of potassium nitrate in the obtained solution.

414. The osmotic pressure of a $0.1\ N$ solution of potassium carbonate K_2CO_3 is 2.69 atm at $0°C$. Determine the apparent degree of ionization of the substance in the solution.

415. A solution containing 8 g of aluminium sulphate $Al_2(SO_4)_3$ in 25 g of water freezes at $-4.46°C$. Determine the apparent degree of ionization of the salt in this solution.

416. A solution containing 0.53 g of sodium carbonate Na_2CO_3 in 200 g of water freezes at $-0.13°C$. Calculate the apparent degree of ionization of Na_2CO_3 in this solution.

417. Determine the apparent degree of ionization of potassium sulphate K_2SO_4 in a solution containing 8.7 g of the salt in 100 g of water, knowing that the solution freezes at $-1.83°C$.

418. When 12 g of sodium hydroxide were dissolved in 100 g of water, the boiling point was raised $2.65°C$. Determine the apparent degree of ionization of sodium hydroxide in the solution.

419. There are solutions containing in equal quantities of water (a) 0.5 mole of sugar and (b) 0.2 mole of calcium chloride. Both solutions freeze at the same temperature. What is the apparent degree of ionization of calcium chloride in the solution?

420. The vapour pressure of a solution containing 0.05 mole of sodium sulphate in 450 g of water is 756.2 mm Hg at 100°C. What is the apparent degree of ionization of Na_2SO_4 in the solution?

421. Determine the isotonic coefficient for a molar solution of an electrolyte knowing that its one litre contains 6.38×10^{23} entities (molecules and ions) of the dissolved substance.

422. Acetic acid CH_3COOH dissociates into H^+ and CH_3COO^- ions. One litre of a 0.01 M solution of acetic acid contains 6.26×10^{21} molecules and ions. Determine the degree of ionization of the acid in the solution.

423. Find the degree of ionization of nitrous acid HNO_2 in a 1 N solution if 1 ml of the solution contains 6.15×10^{20} entities (molecules and ions) of the dissolved substance.

424. What is the osmotic pressure of a 0.1 N solution of potassium chloride, if the apparent degree of ionization of KCl in the solution is 80 per cent?

425. What is the approximate freezing point of a solution containing two moles of common salt per 1,000 g of water, if the apparent degree of ionization of NaCl in this solution is 70 per cent?

426. At 50°C the vapour pressure of water is 92.51 mm Hg. Calculate the vapour pressure of a solution containing one mole of sodium chloride in 1,000 g of water if the apparent degree of ionization of NaCl in this solution is 70 per cent.

427. At what temperature will a solution containing 100 g of sodium chloride in 1,000 g of water start freezing if the apparent degree of ionization of this solution is assumed to be 60 per cent?

428. The osmotic pressure of a 0.04M solution of an electrolyte is 2.15 atm at 0°C and the apparent degree of ionization is 70 per cent. Into how many ions will the electrolyte molecule dissociate?

429. How many moles of a nonelectrolyte does one litre of a solution contain if its osmotic pressure is the same as that of a 1M solution of nitric acid, whose apparent degree of dissociation in this solution is 80 per cent?

430. How many entities (molecules and ions) of the dissolved substance are contained in one millilitre of a 0.1 *M* solution of hydrogen fluoride if the degree of ionization of HF in the solution is 15 per cent?

2. Concentration of Ions and Ionic Equilibrium

Concentration of ions in a solution would be usually expressed by the number of gram-ions per litre of a solution.

A gram-ion is a quantity of an ion, mass of which in grams is numerically equal to its mass in carbon units. For example, a gram-ion of hydrogen is 1 g (more exactly, 1.008 g), a gram-ion of sodium 23 g, a gram-ion of OH^- 17 g, a gram-ion of NO_3^- 62 g and so on.

The concentration of ions depends on the total concentration of a dissolved electrolyte, as well as on the degree of its ionization in a given solution.

Example 1. Calculate the concentration of H^+ and CH_3COO^- ions in a 0.1 *M* solution of acetic acid knowing that the degree of its ionization in the solution is 1.3 per cent.

Solution. In accordance with the conditions of the problem, the total concentration of acetic acid is 0.1 mole/litre. Of this quantity 0.013 mole has dissociated into ions. Since each molecule of acetic acid CH_3COOH dissociates into one H^+ ion and one CH_3COO^- ion, it follows that concentrations of both ions will be equal:

$$[H^+] = [CH_3COO^-] = 0.1 \times 0.013 = 1.3 \times 10^{-3} \text{ g-ion/litre}$$

Example 2. Determine the molar concentration of a solution of ferric chloride $FeCl_3$ in which the concentration of chloride ions is 0.6 g-ion/litre.

Solution. During dissociation of ferric chloride, each molecule of $FeCl_3$ yields three chloride ions. It follows therefore that in complete dissociation of the dissolved salt, the number of chloride gram-ions in the solution will be three times as great as the number of dissolved gram-molecules of ferric chloride, that is the concentration of chloride ions expressed in gram-ions will be three times that of the concentration of $FeCl_3$ expressed in moles.

Example 3. What are concentrations of Ba^{2+} and Cl^- ions in a 0.1 *N* solution of barium chloride $BaCl_2$ if its degree of ionization in the solution is 72 per cent?

Solution. In accordance with the conditions of the problem, the solution contains 0.1 gram-equivalent of barium chloride per litre, which makes 0.05 mole of $BaCl_2$. Of this quantity a part equal to 0.72, that is $0.05 \times 0.72 = 0.036$ mole, was ionized. Since each dissociated molecule of barium chloride yields one barium ion and two chloride ions, the sought concentrations of the ions will obviously be

$$[Ba^{2+}] = 0.036 \text{ g-ion/litre}$$
$$[Cl^-] = 0.036 \times 2 = 0.072 \text{ g-ion/litre}$$

From the above example it follows that the concentration C' of an electrolyte ion can be determined by multiplying the total concentration of the electrolyte C by the degree of its ionization α and the number n which shows how many of these ions are formed from one molecule of the electrolyte:

$$C' = C \cdot \alpha \cdot n$$

It should be noted, however, that the magnitudes thus calculated express the actual concentrations of ions only in weak electrolytes. Strong electrolytes, as has already been noted above, dissociate completely in solutions, and the experimentally found degree of ionization is only apparent. Therefore, concentration of ions calculated from the apparent degree of ionization does not express the actual magnitude of the concentration. This concentration is called 'effective'.

In weak electrolytes the ionization of molecules is a reversible process. Therefore in a solution of a weak electrolyte a state of equilibrium is established between ions and undissociated molecules which obeys the law of acting masses. If we represent the ionization of a weak electrolyte dissociating into two ions by the equation

$$XY \rightleftarrows X^+ + Y^-$$

the following will be true:

$$\frac{[X^+] \cdot [Y^-]}{[XY]} = K$$

The equilibrium constant K is in this case called the *ionization* or *dissociation constant* of an electrolyte.

In the case with weak polybasic acids which ionize by steps, each step of ionization is characterized by its own constant. For example, for sulphurous acid, which ionizes according to the equations

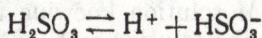

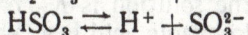

$$H_2SO_3 \rightleftharpoons H^+ + HSO_3^-$$
$$HSO_3^- \rightleftharpoons H^+ + SO_3^{2-}$$

we have

$$K_1 = \frac{[H^+] \cdot [HSO_3^-]}{[H_2SO_3]} = 1.7 \times 10^{-2}$$

$$K_2 = \frac{[H^+] \cdot [SO_3^{2-}]}{[HSO_3^-]} = 1.0 \times 10^{-7}$$

By denoting the total concentration of the dissolved electrolyte by C, the degree of its ionization by α and expressing with their use the concentrations of undissociated molecules and of ions, we have the following:

$$[X^+] = [Y^-] = C\alpha, \qquad [XY] = C(1-\alpha)$$
$$K = \frac{C\alpha \cdot C\alpha}{C(1-\alpha)} = \frac{\alpha^2}{1-\alpha} C$$

This formula can be used for calculating the ionization constants of weak electrolytes if their degree of ionization for a certain concentration is known. On the other hand, if the ionization constant of an electrolyte is known, the degree of ionization can easily be computed at any definite concentration.

If the degree of ionization of an electrolyte is very small, the value of $1-\alpha$ can very well be taken in computations as 1 to simplify the formula as this:

$$K = \alpha^2 C$$

whence

$$\alpha = \sqrt{\frac{K}{C}}$$

Example 4. What is the ionization constant of acetic acid if 1.32 per cent of the acid are ionized in a 0.1 M solution?

Solution. Since the degree of ionization of acetic acid is small, the simplified formula $K = \alpha^2 C$ can be used:

$$K = 0.0132^2 \times 0.1 = 0.0000174 \quad \text{or} \quad 1.74 \times 10^{-5}$$

A more exact calculation yields

$$K = \frac{\alpha^2 C}{1-\alpha} = \frac{(0.0132)^2 \times 0.1}{1-0.0132} = 1.76 \times 10^{-5}$$

Example 5. Determine the degree of ionization of hydrocyanic acid in a 0.05 M solution knowing that the ionization constant of HCN in the solution is $K = 7 \times 10^{-10}$.

Solution. Since hydrocyanic acid is a very weak electrolyte, by applying the formula $K = \alpha^2 C$ we get:

$$7 \times 10^{-10} = \alpha^2 \cdot 0.05$$

whence

$$\alpha = \sqrt{\frac{7 \times 10^{-10}}{0.05}} = 1.18 \times 10^{-4} \quad \text{or} \quad 0.018 \text{ per cent}$$

Example 6. What is the concentration of hydrogen ions in a 0.1 M solution of nitrous acid HNO_2 whose ionization constant is 5×10^{-4}?

Solution. The ionization constant of nitrous acid is expressed by the formula

$$K = \frac{[H^+] \cdot [NO_2^-]}{[HNO_2]}$$

Let the sought concentration of hydrogen ions be x, then

$$[H^+] = [NO_2^-] = x, \qquad [HNO_2] = 0.1 - x$$

$$5 \times 10^{-4} = \frac{x^2}{0.1 - x}$$

Since the magnitude of x is insignificant, it can be omitted in the denominator. Then

$$5 \times 10^{-4} = \frac{x^2}{0.1}$$

$$x = \sqrt{0.5 \times 10^{-4}} = \sqrt{50 \times 10^{-6}} \approx 7 \times 10^{-3} = 0.007 \text{ g-ion/litre}$$

The problem can also be solved by another method: the degree of ionization of nitrous acid in a 0.1 M solution is first found and the concentration of hydrogen ions is then calculated from it:

$$K = \alpha^2 C = 5 \times 10^{-4}$$

$$\alpha = \sqrt{\frac{5 \times 10^{-4}}{0.1}} = 0.07$$

$$[H^+] = 0.1 \times 0.07 = 0.007 \text{ g-ion/litre}$$

If into a solution of a weak electrolyte are introduced ions similar to either of the electrolyte ions, the equilibrium in the solution will be upset and according to the law of acting masses it will be shifted in the direction of formation of undissociated molecules. For example, if to a solution of acetic acid its salt is added, the concentration of CH_3COO^- ions will increase and the equilibrium

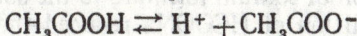

$$CH_3COOH \rightleftarrows H^+ + CH_3COO^-$$

will shift to the left, that is the concentration of undissociated molecules of CH_3COOH will grow, whereas the concentration of hydrogen ions will be reduced. Thus, introduction of similar ions into a solution of a weak electrolyte reduces the degree of its ionization and the concentration of the other ion.

Example 7. The concentration of hydrogen ions in a 0.2 M solution of monobasic formic acid $HCOOH$ is 6.4×10^{-3}. The dissociation constant of formic acid is $K = 2 \times 10^{-4}$. How will the concentration of hydrogen ions decrease, if sodium formiate $HCOONa$ is added to the solution in the amount which adjusts the salt concentration to 1 mole/litre? The apparent degree of ionization of $HCOONa$ in this solution is 0.75.

Solution. Let the sought concentration of hydrogen ions be x. Then the concentration of undissociated molecules is $0.2 - x$, while that of $HCOO^-$ ions will be the sum of two values, viz., concentration of $HCOO^-$ due to the dissociation of the formic acid, and concentration of $HCOO^-$ due to the dissociation of the salt added. The former value will evidently be equal to x, and the latter 0.75. Thus, the concentration of $HCOO^-$ ions will be $0.75 + x$. By substituting the values of the concentrations into the formula of the equilibrium constant we get

$$\frac{x(0.75 + x)}{0.2 - x} = 2 \times 10^{-4}$$

The degree of ionization of formic acid is small and the presence of the salt of this acid will reduce it even more. Therefore, the magnitude of x is very small as compared with 0.75 and 0.2 and it may be disregarded in expressing the concentrations of $HCOO^-$ ions and $HCOOH$ molecules. Then the following equation will be true:

$$\frac{0.75x}{0.2} = 2 \times 10^{-4}$$

whence

$$x = \frac{0.4 \times 10^{-4}}{0.75} = 5.3 \times 10^{-5} \text{ g-ion/litre}$$

By comparing the concentrations of hydrogen ions before and after addition of sodium formiate to the solution, one can see that the addition of HCOONa reduced the hydrogen ion concentration $\frac{6.4 \times 10^{-3}}{5.3 \times 10^{-5}} = 121$ times.

PROBLEMS

431. Determine the concentrations of K^+ and SO_4^{2-} ions in a 0.1 N solution of potassium sulphate K_2SO_4 assuming that the ionization of the salt is complete.

432. Determine the effective concentration of hydroxyl ions in a 0.2 N solution of sodium hydroxide, if the apparent degree of ionization of NaOH is 90 per cent.

433. Determine the effective concentrations of chloride and ferric ions in a 0.1 M solution of ferric chloride $FeCl_3$, knowing that the apparent degree of ionization of the salt in the solution is 65 per cent.

434. Express in gram-ions per litre and in grams per litre the effective concentrations of calcium and chloride ions in a 0.25N solution of calcium chloride, if the apparent degree of ionization of the salt is 72 per cent.

435. Determine the molar concentration of a solution of nitric acid in which the effective concentration of hydrogen ions is 0.294 g-ion/litre and the apparent degree of ionization is 84 per cent.

436. Calculate the concentration of cupric nitrate $Cu(NO_3)_2$ solution knowing that the apparent degree of ionization of the salt in this solution is 64 per cent and the effective concentration of NO_3^- ions is 0.16 g-ion/litre.

437. The solubility of potassium chlorate $KClO_3$ is 0.52 mole/litre at 18°C. What are the effective concentrations of K^+ and ClO_3^- ions in a saturated solution, if 75 per cent of the salt have dissociated?

438. Determine the normality of a solution of copper sulphate $CuSO_4$ in which the effective concentration of copper ions is 0.02 g-ion/litre and the apparent degree of ionization of the salt is 40 per cent.

439. Find the effective concentrations of magnesium and chloride ions in a 20 per cent solution of magnesium chloride (density 1.16 g/cu cm) knowing that 54 per cent of the salt have dissociated.

440. A solution containing 0.5 mole of NaCl, 0.16 mole of KCl and 0.24 mole of K_2SO_4 in 1 litre was needed for an experiment. NaCl, KCl and Na_2SO_4 were only available, but the assistant prepared the required solution. How did he solve the problem?

441. Prove that a solution containing 0.6 mole of $CuSO_4$, 0.25 mole of $Cu(NO_3)_2$ and 0.35 mole of K_2SO_4 in one litre is similar to a solution containing 0.85 mole of $CuSO_4$, 0.5 mole of KNO_3 and 0.1 mole of K_2SO_4 in one litre.

442. The ionization constant of monobasic butyric acid C_3H_7COOH is 1.5×10^{-5}. What is the degree of its ionization in a 0.005 M solution?

443. Determine the degree of ionization of hypochlorous acid HClO in a 0.2N solution if its ionization constant is 4×10^{-8}.

444. The degree of ionization of formic acid HCOOH in a 0.2 N solution is 3.2 per cent. Determine the ionization constant of formic acid.

445. The degree of the first ionization of carbonic acid ($H_2CO_3 \rightleftarrows H^+ + HCO_3^-$) in a 0.1 N solution is 0.173 per cent. Calculate the ionization constant of this step.

446. The apparent degree of ionization of hydrochloric acid in a 1 N solution is 78 per cent and in a decinormal solution 92 per cent. Prove that the ratio $\frac{[H^+] \cdot [Cl^-]}{[HCl]}$ is not constant for hydrochloric acid and depends on the concentration of the solution.

447. At what concentration of a solution will the degree of ionization of nitrous acid HNO_2 be equal to 20 per cent, if the ionization constant of the acid is 5×10^{-4}?

448. The degree of ionization of acetic acid in a 0.1 N solution is 1.32 per cent, and that of hydrochloric acid at the same concentration is 92 per cent. At what concentration will the degree of ionization of acetic acid also be 92%?

449. How many millilitres of water must be added to 300 ml of 0.2 M solution of acetic acid ($K = 1.8 \times 10^{-5}$) to double the degree of ionization?

450. What is the concentration of hydrogen ions in an aqueous solution of formic acid ($K = 2 \times 10^{-4}$) if its degree of ionization is 5 per cent?

451. Calculate the concentration of hydrogen ions in a 0.02 M solution of sulphuric acid taking into account only the first ionization for which the constant is 1.7×10^{-2}.

452. How will the concentration of hydrogen ions decrease if 0.1 mole of sodium acetate is added to one litre of a 0.2 M solution of acetic acid? The degree of ionization of sodium acetate at this dilution is 80 per cent. Assume the ionization constant of acetic acid to be 1.8×10^{-5}.

453. Calculate the concentration of CH_3COO^- ions in a solution whose one litre contains 1 mole of acetic acid and 0.1 mole of HCl, assuming that the dissociation of HCl is complete. Take the ionization constant of acetic acid to be equal to 1.8×10^{-5}.

3. Solubility Product

In any saturated solution containing a solid precipitate of the solute the equilibrium is set up between the dissolved and undissolved solute. If the dissolved solute is a strong slightly soluble electrolyte, for example slightly soluble salt, the saturated solution contains only separate ions of this salt which are in equilibrium with the precipitate. For example:

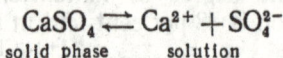

$$\underset{\text{solid phase}}{CaSO_4} \rightleftarrows \underset{\text{solution}}{Ca^{2+} + SO_4^{2-}}$$

In a saturated solution of a slightly soluble electrolyte the product of the concentrations of its ions is constant at any definite temperature. This value is known as the **solubility product** of the electrolyte and is denoted by SP (or by the letter L). For example, the solubility product of $CaSO_4$

$$SP_{CaSO_4} = [Ca^{2+}] \cdot [SO_4^{2-}]$$

In cases where two or more identical ions are produced in dissociation of an electrolyte, the concentration of these ions should be raised to the corresponding power when calculating the solubility product. For example

$$SP_{Ca_3(PO_4)_2} = [Ca^{2+}]^3 \cdot [PO_4^{3-}]^2$$

The solubility product characterizes the solubility of a solid electrolyte at a given temperature. If there are two similar salts (for example, $CaSO_4$ and $BaSO_4$) the solubility is greater in that salt whose solubility product is greater.

The concentration of any ion in a saturated solution can be varied, but this will induce immediate response in the concentration of the other ion to ensure the constancy of the solubility product. Therefore, if a certain amount of either ion is introduced into a saturated solution of an electrolyte, the concentration of the other ion must change accordingly and a part of the dissolved substance will precipitate.

Thus, the solubility of an electrolyte decreases when the same ions are added to its solution.

The solubility product value makes it possible to solve many important problems in chemistry.

Example 1. The solubility product of lead iodide PbI_2 at room temperature is 1.4×10^{-8}. Calculate the solubility of this salt at the same temperature and also concentration of either ion in a saturated solution.

Solution. PbI_2 ionizes in solution into Pb^{2+} and I^- ions. Since each molecule of lead iodide yields one lead ion and two iodide ions, the concentration of lead ions expressed in gram-ions per litre is equal to the total concentration of the dissolved salt, and the concentration of iodide ions is twice as great.

Let the molar concentration of the saturated solution of lead iodide be x, then

$$[PbI_2] = x, \qquad [Pb^{2+}] = x, \qquad [I^-] = 2x$$

In accordance with the conditions of the problem

$$SP_{PbI_2} = [Pb^{2+}] \cdot [I^-]^2 = 1.4 \times 10^{-8}$$

By substituting into the equation we have the following:

$$4x^3 = 1.4 \times 10^{-8}, \qquad x^3 = 0.35 \times 10^{-8} = 3.5 \times 10^{-9},$$
$$x = \sqrt[3]{3.5 \times 10^{-9}} = 1.5 \times 10^{-3}$$

The solubility of a substance at a given temperature is measured by the concentration of its saturated solution. Hence, the solubility of PbI_2 is 1.5×10^{-3} mole/litre; $[Pb^{2+}] = 1.5 \times 10^{-3}$ g-ion/litre; $[I^-] = 3 \times 10^{-3}$ g-ion/litre.

Example 2. The solubility of magnesium hydroxide $Mg(OH)_2$ at 18°C is 2×10^{-4} mole/litre. What is the solubility product of this substance?

Solution. The total concentration of the saturated solution of magnesium hydroxide is 2×10^{-4} mole/litre. During dissociation one molecule of $Mg(OH)_2$ yields one magnesium ion and two hydroxyl ions. Hence:

$[Mg^{2+}] = 2 \times 10^{-4}$ g-ion/litre, $[OH^-] = 4 \times 10^{-4}$ g-ion/litre

The solubility product of $Mg(OH)_2$ is

$$SP_{Mg(OH)_2} = [Mg^{2+}] \cdot [OH^-]^2$$

By substituting into this equation the values of ion concentrations we get

$$SP_{Mg(OH)_2} = 2 \times 10^{-4}(4 \times 10^{-4})^2 = 3.2 \times 10^{-11}$$

Example 3. How does the solubility of calcium oxalate CaC_2O_4 in a 0.05 M solution of ammonium oxalate $(NH_4)_2C_2O_4$ differ from that in pure water if the apparent degree of ionization of ammonium oxalate in these conditions is 70 per cent and the solubility product of calcium oxalate is 3.8×10^{-9}?

Solution. Calculate first the solubility of calcium oxalate CaC_2O_4 in pure water. Let the molar concentration of the saturated solution of this salt be x. Since a molecule of calcium oxalate dissociates into two ions, the concentration of either ion will be x. Hence, taking into account the magnitude of the solubility product of CaC_2O_4 we get:

$$x^2 = 3.8 \times 10^{-9}, \qquad x = \sqrt{3.8 \times 10^{-9}} = 6.2 \times 10^{-5}$$

It follows therefore that the solubility of calcium oxalate in pure water is 6.2×10^{-5} mole/litre.

Now the solubility of the same salt in a 0.05 M solution of ammonium oxalate should be found. Let it be denoted by y. The concentration of Ca^{2+} ions in the solution will also be y. As to $C_2O_4^{2-}$ ions, their concentration will be the sum of y and the concentration of $C_2O_4^{2-}$ ions in the 0.05 M solution of ammonium oxalate which (in accordance with the conditions of the problem) is $0.05 \times 0.7 = 0.035$ g-ion/litre. Since the magnitude of y is very small compared with 0.035, it may be ignored in the calculations and the concentration of $C_2O_4^{2-}$ ions can be assumed to be 0.035 g-ion/litre.

Since the product of concentrations of $C_2O_4^{2-}$ ions and Ca^{2+} ions must be equal to the solubility product of calcium oxalate it follows that

$$0.035 \cdot y = 3.8 \times 10^{-9}, \qquad y = 1.09 \times 10^{-7}$$

It follows therefore that one litre of the saturated solution of calcium oxalate will contain 1.09×10^{-7} mole of calcium oxalate. If we compare now this value with that found earlier (6.2×10^{-5}) we can arrive at the conclusion that the solubility of the calcium oxalate has reduced $\frac{6.2 \times 10^{-5}}{1.09 \times 10^{-7}}$ times, that is approximately 570 times.

Example 4. Equal volumes of $0.02N$ solutions of calcium chloride and sodium sulphate are mixed. Will calcium sulphate $CaSO_4$ precipitate? The solubility product of calcium sulphate is 2.3×10^{-4}.

Solution. Since the volume of the obtained mixture is twice as great as the volume of either component, the concentration of each ion after mixing will decrease two times. Assuming that the salts have ionized completely, we can write

$$[CaCl_2] = [Ca^{2+}] = 0.01 \times 0.5 = 5 \times 10^{-3}$$
$$[Na_2SO_4] = [SO_4^{2-}] = 0.01 \times 0.5 = 5 \times 10^{-3}$$

whence

$$[Ca^{2+}] \cdot [SO_4^{2-}] = (5 \times 10^{-3})^2 = 2.5 \times 10^{-5}$$

This calculation shows that the product of concentrations of Ca^{2+} and SO_4^{2-} ions in the obtained mixture is less than the value of the solubility product of calcium sulphate. Hence, with respect to calcium sulphate the solution will be unsaturated and the salt will not precipitate.

PROBLEMS

454. The solubility of lead iodide PbI_2 at 18°C is $1.5 \times \times 10^{-3}$ mole/litre. Calculate the concentrations of lead and iodide ions in a saturated solution of lead iodide assuming that the dissociation is complete.

455. The solubility of calcium carbonate $CaCO_3$ at 18°C is 1.3×10^{-4} mole/litre. Calculate the solubility product of this salt.

456. The solubility of lead bromide $PbBr_2$ at 18°C is $2.7\times$ $\times 10^{-2}$ mole/litre. Calculate the solubility product of this salt.

457. In 500 ml of water at 18°C, 0.0165 g of silver chromate Ag_2CrO_4 is dissolved. What is the solubility product of this salt?

458. The concentration of fluoride ions in a saturated at 18°C solution of calcium fluoride CaF_2 is 4×10^{-4} g-ion/litre. Find the solubility product of calcium fluoride.

459. In order to dissolve 1 g of lead iodide PbI_2 at 18°C, 1,470 ml of water are required. What is the solubility product of this salt?

460. The solubility product of calcium carbonate is $1.7\times$ $\times 10^{-8}$. How many grams of $CaCO_3$ are contained in one litre of the saturated solution?

461. The solubility product of silver bromide AgBr is $3.6\times$ $\times 10^{-13}$. How many grams of silver in the form of ions does one litre of the saturated solution of this salt contain?

462. The solubility product of silver sulphate Ag_2SO_4 is 7×10^{-5}. Find the solubility of the salt and express it in moles per litre and in grams per litre.

463. How many litres of water are required to dissolve 1 g of barium carbonate at room temperature? The solubility product of $BaCO_3$ is 1.9×10^{-9}.

464. How many grams of lead sulphate can be dissolved at room temperature in one litre of water, if the solubility product of $PbSO_4$ is 2.3×10^{-8}?

465. The solubility product of strontium sulphate $SrSO_4$ is 3.6×10^{-7}. If we mix equal volumes of 0.002 N solutions of strontium chloride $SrCl_2$ and potassium sulphate K_2SO_4 will strontium sulphate precipitate?

466. How much will the concentration of silver ions in a saturated solution of AgCl decrease if hydrochloric acid is added to it in amount which will bring the concentration of HCl in the solution to 0.03 mole/litre? The solubility product of the silver chloride is 1.2×10^{-10}.

467. The solubility product of silver sulphate Ag_2SO_4 is 7×10^{-5}. If we add to a $0.02N$ solution of silver nitrate $AgNO_3$ an equal volume of $1N$ solution of sulphuric acid, will the salt precipitate?

468. The solubility product of lead chloride $PbCl_2$ is

2.3×10^{-4}. Will lead chloride precipitate if to a $0.1N$ solution of lead nitrate an equal volume of $0.4N$ solution of sodium chloride is added?

4. Ionic Reactions and Equations

The reactions in solutions of electrolytes would be usually the interaction between ions of the dissolved substances. Although the majority of these reactions belong to the so-called exchange reaction, from the viewpoint of the ionic theory the essence of these reactions is the combination of ions into molecules of new substances. Ions, however, can unite only on condition that the new substance is slightly soluble or is a weak electrolyte. The requisite condition for the exchange reaction between electrolytes in solutions is therefore formation of sparingly soluble or weakly dissociating substances.

Example 1. The reaction between calcium chloride and sodium carbonate in an aqueous solution is expressed by the molecular equation

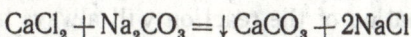

$$CaCl_2 + Na_2CO_3 = \downarrow CaCO_3 + 2NaCl$$

Being a solution of a strong electrolyte, calcium chloride solution contains only Ca^{2+} and Cl^- ions. Likewise, the sodium carbonate dissociates completely in the solution into Na^+ and CO_3^{2-} ions. When the solutions of $CaCl_2$ and Na_2CO_3 are mixed together, Ca^{2+} and CO_3^{2-} ions combine to form a practically insoluble in water calcium carbonate which precipitates. The molecules of NaCl indicated in the equation are not actually formed since sodium chloride is a strong electrolyte and it can be present in solution only in the form of ions. It follows therefore that only Ca^{2+} and CO_3^{2-} ions interact chemically, whereas Na^+ and Cl^- ions remain free after mixing, in other words, they did not take part in the reaction. Taking this into account, the reaction can be depicted in the following way:

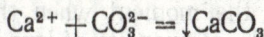

$$Ca^{2+} + CO_3^{2-} \rightleftharpoons \downarrow CaCO_3$$

This equation is called the *ionic equation* of the reaction. It is simpler than the molecular equation and describes more exactly the essence of the reaction.

Example 2. A reaction which takes place during mixing of sodium hydroxide and hydrochloric acid solutions (neutralization reaction) is expressed by the molecular equation

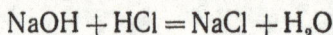

$$NaOH + HCl = NaCl + H_2O$$

As in the previous case, the solutions of the starting substances NaOH and HCl contain only ions of these substances. Since sodium chloride is a strong electrolyte, only hydrogen and hydroxyl ions will combine on mixing the solutions, water, a weak electrolyte, being formed as a result. The formation of molecules of water characterizes the course of the reaction, whose essence can thus be expressed by the ionic equation

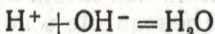

$$H^+ + OH^- = H_2O$$

Any other reaction of neutralization of a strong acid with a strong base will evidently be expressed by a similar equation.

* * *

If the reactants are strong and readily soluble electrolytes, the ionic equation of the reaction will always have the form as in the above examples, namely, the interacting substances are placed in the left part of the equation, whereas the resultant slightly dissociating or sparingly soluble substances are placed in its right part. Ions can be regarded as the starting substances in these reactions and molecules of a weak electrolyte or precipitates of sparingly soluble substances as the products of the reaction.

The reactions are somewhat more complicated in cases where one of the reactants is a weak or a slightly soluble electrolyte. Although in these reactions the particles which react directly with each other are ions too, and the essence of the reaction consists in their combining, however in respect of a weak or slightly soluble electrolyte, its molecules or crystals rather than ions should be considered to be the initial substance, since at the moment when the reaction begins, there is only insignificant quantity of ions in the solution whereas the rest of them are formed in the reaction. In this connection the ionic equations of these reactions have a different appearance.

Example 3. During mixing a weak hydrocyanic acid HCN with a solution of sodium hydroxide a reaction takes place whose molecular equation is

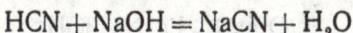

$$HCN + NaOH = NaCN + H_2O$$

The solution of hydrocyanic acid contains mainly molecules of HCN which are in equilibrium with a small number of H^+ and CN^- ions:

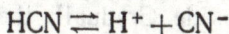

$$HCN \rightleftharpoons H^+ + CN^-$$

After the alkali is added to this solution its hydroxyl groups start reacting with hydrogen ions of the acid to form molecules of water. But the loss of hydrogen ions from the solution upsets the equilibrium and new molecules of the hydrocyanic acid dissociate. Thus the entire hydrocyanic acid gradually dissociates to release hydrogen ions which combine with the hydroxyl ions to form molecules of water.

In this reaction the starting substances are molecules of hydrocyanic acid, which ionize completely only in the course of the reaction, and hydroxyl ions of the alkali which are present in the solution in the requisite quantities from the very beginning of the reaction. In addition to molecules of water, the products of the reaction are also CN^- ions which are released from the molecules of the hydrocyanic acid. Therefore the ionic equation of this reaction can be written as this:

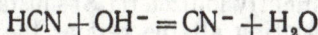

$$HCN + OH^- = CN^- + H_2O$$

The ionic equation has a similar appearance in the case where one of the reactants is sparingly soluble in water and in the beginning of the reaction is present in the form of a precipitate.

Example 4. The reaction of ferric hydroxide $Fe(OH)_3$ dissolution in hydrochloric acid is expressed by the molecular equation:

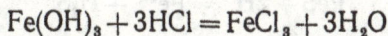

$$Fe(OH)_3 + 3HCl = FeCl_3 + 3H_2O$$

The reaction is due to the formation of molecules of water from hydrogen and hydroxyl ions. While all hydrogen ions are present in the solution from the start hydroxyl ions are formed in sufficient quantities only in the course of the reaction from the ferric hydroxide which is almost insoluble in water. By the moment when the reaction starts, the ferric hydroxide is in the form of a precipitate in equilibrium with insignificant quantities of Fe^{3+} and OH^- ions in the solution:

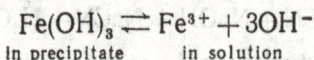

$$Fe(OH)_3 \rightleftharpoons Fe^{3+} + 3OH^-$$
In precipitate   in solution

As the acid is added, hydroxyl ions are bonded with hydrogen ions to upset the equilibrium and to dissolve new quantities of the ferric hydroxide. Gradually the entire precipitate is dissolved and simultaneously Fe^{3+} ions are released. Hence, the starting substances in the reaction in question are molecules of ferric hydroxide and hydrogen ions, whereas the products of the reaction are ferric ions and molecules of water. Therefore, the ionic equation of the reaction will be as this:

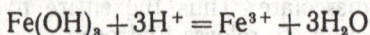

$$Fe(OH)_3 + 3H^+ = Fe^{3+} + 3H_2O$$

* * *

On the basis of all that was said about ionic equations the following order of deriving them may be recommended:

1. Write the equation of the reaction in its molecular form, for example, the reaction of FeS dissolution in hydrochloric acid:

$$FeS + 2HCl = FeCl_2 + H_2S$$

2. Rewrite the equation leaving the practically insoluble in water (FeS) or slightly ionized (H_2S) substances in the molecular form, and writing all the rest as the ions into which they break up:

$$FeS + 2H^+ + 2Cl^- = Fe^{2+} + 2Cl^- + H_2S$$

3. Cancel the ions that do not take part in the reaction, that is those found in equal numbers on the left and the right sides of the equation (in this particular case Cl^- ions):

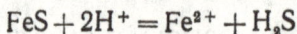

$$FeS + 2H^+ = Fe^{2+} + H_2S$$

PROBLEMS

In solving problems of this section, consult the Table of solubility of most important salts given in Chapter VI, Sec. 7.

469. Represent the following reactions in the ionic form:

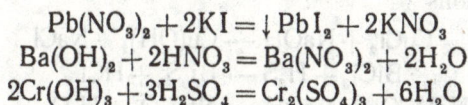

$$Pb(NO_3)_2 + 2KI = \downarrow PbI_2 + 2KNO_3$$
$$Ba(OH)_2 + 2HNO_3 = Ba(NO_3)_2 + 2H_2O$$
$$2Cr(OH)_3 + 3H_2SO_4 = Cr_2(SO_4)_3 + 6H_2O$$

470. Write the ionic equations for the following reactions:

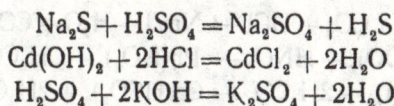

$$Na_2S + H_2SO_4 = Na_2SO_4 + H_2S$$
$$Cd(OH)_2 + 2HCl = CdCl_2 + 2H_2O$$
$$H_2SO_4 + 2KOH = K_2SO_4 + 2H_2O$$

471. Write out molecular and ionic equations for the following reactions: (a) reaction between aluminium chloride $AlCl_3$ and silver sulphate Ag_2SO_4; (b) reaction of neutralization of a weak acetic acid CH_3COOH with an alkali; (c) action of hydrogen sulphide H_2S on a solution of nickel chloride $NiCl_2$; (d) dissolution of magnesium hydroxide $Mg(OH)_2$ in nitric acid.

472. Derive ionic equations for the following reactions: (a) dissolution of ferrous sulphide FeS and ferric hydroxide $Fe(OH)_3$ in sulphuric acid; (b) neutralization of a weak nitrous acid HNO_2 with sodium hydroxide; (c) action of an alkali on a solution of copper sulphate $CuSO_4$.

473. Mix in pairs the following solutions: Na_3PO_4, $CuSO_4$, $ZnCl_2$ and K_2S (six combinations altogether). Indicate combinations where a reaction will take place. Derive their ionic equations.

474. Write ionic equations for the reactions of dissolution of zinc sulphide ZnS and zinc hydroxide $Zn(OH)_2$ in hydrochloric acid and the reactions which take place in mixing a solution of barium hydroxide $Ba(OH)_2$ with solutions of ferric chloride $FeCl_3$ and of magnesium sulphate $MgSO_4$.

475. Mix in pairs the solutions of $Pb(NO_3)_2$, Na_2S, $CuSO_4$, and KCl (a total of six combinations). Indicate combinations where a reaction will take place and write their molecular and ionic equations.

476. Derive molecular and ionic equations for the reactions in solutions between hydrogen sulphide and nickel sulphate, potassium bicarbonate and potassium hydroxide, ferric hydroxide and nitric acid.

477. Derive molecular and ionic equations for the following reactions:

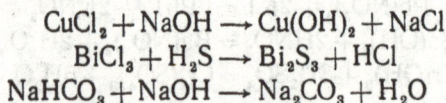

$$CuCl_2 + NaOH \longrightarrow Cu(OH)_2 + NaCl$$
$$BiCl_3 + H_2S \longrightarrow Bi_2S_3 + HCl$$
$$NaHCO_3 + NaOH \longrightarrow Na_2CO_3 + H_2O$$

478. Derive ionic equations for the following reactions:

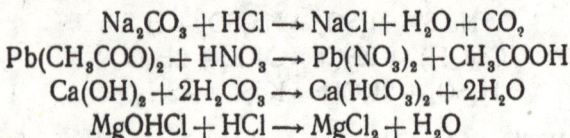

$$Na_2CO_3 + HCl \longrightarrow NaCl + H_2O + CO_2$$
$$Pb(CH_3COO)_2 + HNO_3 \longrightarrow Pb(NO_3)_2 + CH_3COOH$$
$$Ca(OH)_2 + 2H_2CO_3 \longrightarrow Ca(HCO_3)_2 + 2H_2O$$
$$MgOHCl + HCl \longrightarrow MgCl_2 + H_2O$$

479. Derive molecular equations for the reactions in solutions between calcium chloride and ammonium carbonate, calcium sulphate and soda, calcium nitrate and soda, calcium bromide and potash. What is the essence of these reactions? Can they be expressed by a single ionic equation? What is this equation?

480. Derive ionic equations for the reactions in solutions between barium carbonate and sulphuric acid, water-insoluble basic bismuth nitrate and hydrochloric acid, barium hydroxide and nickel chloride.

481. Select substances whose aqueous solutions would react according to the following ionic equations:

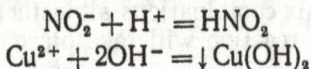

$$NO_2^- + H^+ = HNO_2$$
$$Cu^{2+} + 2OH^- = \downarrow Cu(OH)_2$$

Write the corresponding molecular equations.

482. Select molecular equations for the reactions which are expressed by the following ionic equations:

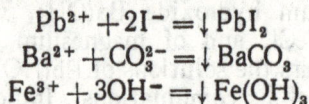

$$Pb^{2+} + 2I^- = \downarrow PbI_2$$
$$Ba^{2+} + CO_3^{2-} = \downarrow BaCO_3$$
$$Fe^{3+} + 3OH^- = \downarrow Fe(OH)_3$$

483. Derive two complete (molecular) equations for each of the following reactions:

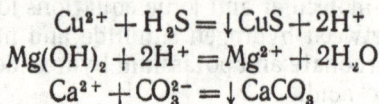

$$Cu^{2+} + H_2S = \downarrow CuS + 2H^+$$
$$Mg(OH)_2 + 2H^+ = Mg^{2+} + 2H_2O$$
$$Ca^{2+} + CO_3^{2-} = \downarrow CaCO_3$$

5. Hydrogen Ion Concentration

In pure water, an insignificant part of H_2O molecules dissociates into ions:

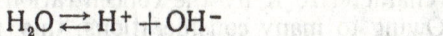

$$H_2O \rightleftarrows H^+ + OH^-$$

Experimental measurements have shown that at room temperature, and more exactly at 24°C, the concentrations of hydrogen and of hydroxyl ions in pure water are 10^{-7} g-ion/litre:

$$[H^+] = [OH^-[= 10^{-7} \text{ g-ion/litre}$$

The product of concentrations of hydrogen and hydroxyl ions of water is called the *ion product of water*. Numerically it is equal to 10^{-14}. The ion product of water is a constant not only for water but also for dilute aqueous solutions of any substance. Let us denote the constant as K_{H_2O}, then:

$$K_{H_2O} = [H^+] \cdot [OH^-] = 10^{-14}$$

Strictly speaking, the ion product for water is 10^{-14} only at 24°C. It increases with growing temperature and decreases with lowering temperature. However, for calculations related to room temperatures K_{H_2O} can always be assumed to be 10^{-14}.

Solutions in which the concentrations of hydrogen and hydroxyl ions are the same and equal 10^{-7} gram-ion per litre each are called neutral solutions. In acid solutions the concentration of hydrogen ion is higher and in alkaline solutions, it is the concentration of hydroxyl ion that is higher. But whatever the concentrations of H^+ and OH^- ions in the solution, their product remains constant, 10^{-14}. Therefore, if the concentration of either of the water ions is known, the concentration of the other ion can easily be calculated:

$$[H^+] = \frac{10^{-14}}{[OH^-]}, \qquad [OH^-] = \frac{10^{-14}}{[H^+]}$$

For example, if the concentration of hydrogen ions in a given solution is 2×10^{-5} g-ion/litre, the concentration of hydroxyl ions is 10^{-14} divided by 2×10^{-5}, that is 5×10^{-10} g-ion/litre. If the concentration of hydroxyl ions is 10^{-6} g-ion/litre, the concentration of hydrogen ions is 10^{-8} g-ion/litre. etc.

Thus, the reaction of any aqueous solution, its acidity or alkalinity can quantitatively be characterized by the concentration of only one of the water ions. The general practice is to characterize it by the concentration of hydrogen ion.

Owing to many considerations, this method of expressing the acidity or alkalinity of a solution has given way to another, still more simple and convenient method: instead of indicating the true ion concentration, we indicate the logarithm of the latter with its sign reversed. This value is designated pH. For example, if the concentration of hydrogen ion is 10^{-6} g-ion/litre, then pH=6. If $[H^+]=5\times10^{-5}$, then pH=5, etc.

It is quite evident that mathematically the pH is a common logarithm of the concentration of hydrogen ion with its sign reversed:

$$pH = - \log[H^+]$$

From what has been said it follows that
(1) the pH of a neutral solution is 7;
(2) the pH of an acid solution is less than 7, diminishing with the growing acidity of a solution;
(3) the pH of an alkaline solution is greater than 7.

Once the pH of a solution is known, it is easy to calculate the concentration of hydrogen or hydroxyl ions in the solution. And vice versa, if the concentration of hydrogen or hydroxyl ions is known, one can establish the value of pH of the solution. During calculations, the logarithms are always rounded to 0.01, since we cannot determine the pH values to a greater accuracy.

Example 1. The concentration of hydrogen ion in a solution is 0.004 g-ion/litre. What is the pH of the solution?

Solution. Knowing that $pH=-\log[H^+]$, and rounding the logarithm to 0.01, we get:

$$pH = - \log 0.004 = -\bar{3}.60 = -(-3+0.60) = 2.4$$

Example 2. Determine the concentration of hydrogen ion in a solution whose pH is 4.6.

Solution. In accordance with the conditions of the problem

$$- \log[H^+] = 4.6 \quad \text{or} \quad \log[H^+] = -4.6$$

To determine the concentration of hydrogen ion from these data, it is necessary first to find the number (antilogarithm) from the given logarithm. If we want to use the antilogarithm table, the logarithm should first be rearranged so that only its characteristic is negative, whereas the mantissa is positive. To this end, we shall substract 1 from the characteristic and add 1 to the mantissa:

$$-4.6 = (-4 \ -1) + (-0.6 + 1) = -5 + 0.4 = 5.4$$

Thus

$$\log [H^+] = \overline{5}.4$$

Now we can find the corresponding antilogarithm from the table:

$$[H^+] = 0.000025 = 2.5 \times 10^{-5} \text{ g-ion/litre}$$

Example 3. What is the concentration of hydroxyl ion in a solution whose pH is 10.8?

Solution. The concentration of hydrogen ion in the solution should first be determined:

$$- \log [H^+] = 10.8$$
$$\log [H^+] = -10.8 = \overline{11}.2$$
$$[H^+] = 1.6 \times 10^{-11} \text{ g-ion/litre}$$

Since the product $[H^+] \cdot [OH^-]$ is 10^{-14} we have

$$[OH^-] = \frac{10^{-14}}{1.6 \times 10^{-11}} = 6.25 \times 10^{-4} \text{ g-ion/litre}$$

PROBLEMS

484. Determine the pH of a $0.002N$ solution of nitric acid assuming that the ionization is complete.

485. Determine the pH of a $0.01N$ solution of acetic acid whose degree of ionization in this solution is 4.2 per cent.

486. Determine the pH of a solution containing 0.1 g of NaOH in one litre assuming that the ionization is complete.

487. What is the pH of a solution whose one litre contains 0.0051 g of hydroxyl ion?

488. Determine the concentration of hydrogen and hydroxyl ions in a solution whose pH is 6.2.

489. Calculate the concentration of hydrogen ion and the pH of a $0.5M$ solution of HCl ionized to 85 per cent.

490. How many hydrogen ions does 1 ml of a solution contain at pH 13?

491. How shall the concentration of hydrogen ion be changed in order to increase the pH of a solution by 1?

492. How will the pH of pure water change if 0.001 mole of NaOH is added to one litre?

493. How many millilitres of a $0.1N$ solution of an alkali should be added to 10 ml of a $0.5N$ solution of an acid to adjust its pH to 7?

494. Distilled water exposed to air contains 1.35×10^{-5} mole of CO_2. Calculate the pH of the water taking into account only the first ionization of carbonic acid for which the ionization constant is 3×10^{-7}.

6. Hydrolysis of Salts

The hydrolysis of a salt is its interaction with water as a result of which an acid (or an acid salt) and a base (or a basic salt) are formed.

From the ionic theory viewpoint, hydrolysis is the combination of salt ions with the hydrogen or hydroxyl ions of water (sometimes with both) and formation of undissociated molecules or new weakly dissociating ions.

Since strong acids and bases are practically ionized completely, it is evident that, of salt ions, only those of weak acid radicals and those of metals forming weak bases can unite with water ions. It follows, therefore, that those salts are only ionized which contain the said ions, in other words, salts formed (a) by a weak acid and a strong base (for example, NaCN); (b) by a strong acid and a weak base (for example, $ZnCl_2$); and (c) by a weak acid and a weak base (for example, $Al(CH_3COO)_3$). A salt of a strong acid and a strong base, like NaCl, is not hydrolyzed. Hydrogen or hydroxyl ions of water combine with ions of the dissolved salt to upset the equilibrium between the molecules of water and its ions, and the solution becomes either acid or alkaline.

Hydrolysis is a reversible process. In most cases the hydrolyzed part of salt is so small that the hydrolysis products, even if they are practically insoluble (for example hydroxides of heavy metals or basic salts), do not precipitate and remain in solution.

Rules used in deriving ionic equations for common exchange reactions hold for hydrolysis too.

Let us consider hydrolysis of potassium cyanide KCN as an example.

Hydrolysis of this salt is due to the formation of undissociated molecules of a very weak hydrocyanic acid. The molecular equation of the reaction is:

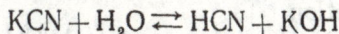

$$KCN + H_2O \rightleftarrows HCN + KOH$$

The ionic equation of the same reaction is

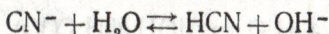

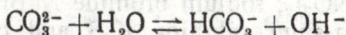

$$CN^- + H_2O \rightleftarrows HCN + OH^-$$

Since excess hydroxyl ions are produced in the hydrolysis the reaction of the potassium cyanide solution is alkaline.

* * *

During hydrolysis of salts containing a cation of a strong base and an anion of a weak dibasic or polybasic acid (for example, Na_2CO_3 or K_3PO_4), acid salts (more exactly, anions of acid salts) rather than free acids are formed as a rule. Basic salts (or cations of basic salts), rather than free bases, are produced likewise due to hydrolysis of salts formed by polyvalent weak bases and strong acids (for example, $FeCl_3$).

Consider a few examples.

1. *Hydrolysis of sodium carbonate* Na_2CO_3. During dissolution of this salt in water, CO_3^{2-} ions are associated with hydrogen ions of water to form almost undissociating HCO_3^- ions. Formation of these ions (not of H_2CO_3 molecules) is explained by that they are ionized less readily than the molecules of carbonic acid. In the molecular form the reaction can be expressed by the following equation:

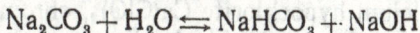

$$Na_2CO_3 + H_2O \rightleftarrows NaHCO_3 + NaOH$$

The ionic equation of the reaction:

$$CO_3^{2-} + H_2O \rightleftarrows HCO_3^- + OH^-$$

The reaction of the solution is on the alkaline side due to the presence of excess hydroxyl ions.

2. *Hydrolysis of zinc chloride* $ZnCl_2$. The reaction is due to the formation of very weakly dissociating $ZnOH^-$ ions.

The molecular equation of this reaction is

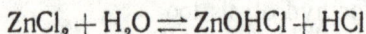

$$ZnCl_2 + H_2O \rightleftharpoons ZnOHCl + HCl$$

The ionic equation of the reaction is

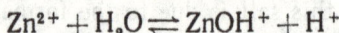

$$Zn^{2+} + H_2O \rightleftharpoons ZnOH^+ + H^+$$

Owing to the presence of excess hydrogen ions the solution has an acid reaction.

3. *Hydrolysis of aluminium acetate* $Al(CH_3COO)_3$. A weak acetic acid and a basic salt $AlOH(CH_3COO)_2$ are formed:

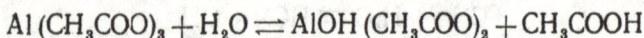

$$Al(CH_3COO)_3 + H_2O \rightleftharpoons AlOH(CH_3COO)_2 + CH_3COOH$$

The ionic form of the equation:

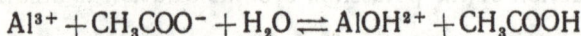

$$Al^{3+} + CH_3COO^- + H_2O \rightleftharpoons AlOH^{2+} + CH_3COOH$$

If the solution is boiled, the hydrolysis is intensified, another basic salt $Al(OH)_2CH_3COO$ precipitates, and the reaction proceeds practically to the end:

$$AlOH(CH_3COO)_2 + H_2O = \downarrow Al(OH)_2CH_3COO + CH_3COOH$$

PROBLEMS

495. Derive molecular and ionic equations for hydrolysis of the following salts:

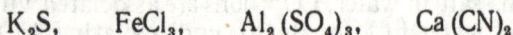

$$K_2S, \quad\quad FeCl_3, \quad\quad Al_2(SO_4)_3, \quad\quad Ca(CN)_2$$

496. Derive molecular and ionic equations for hydrolysis of sodium salts of weak acids: nitrous acid HNO_2 and hydrosulphuric acid H_2S.

497. Derive ionic equations for hydrolysis of copper nitrate $Cu(NO_3)_2$, calcium hypochlorite $Ca(ClO)_2$, chromium trichloride $CrCl_3$ and sodium sulphide Na_2S.

498. Determine the reaction of the following solutions: zinc nitrate $Zn(NO_3)_2$, sodium bromide $NaBr$, barium chloride $BaCl_2$, potassium sulphide K_2S and cupric chloride $CuCl_2$.

Prove your answer by the corresponding ionic equations.

499. What salts of those given below will hydrolyze?

$$CrCl_3, \quad\quad K_2CO_3, \quad\quad NaCN, \quad\quad CaBr_2, \quad\quad NaNO_3, \quad\quad K_2SO_4$$

Express their hydrolysis by ionic equations and indicate the reaction of their solutions.

500. What is the reaction to litmus of the following solutions: sodium cyanide NaCN, copper sulphate $CuSO_4$, potassium nitrate KNO_3, sodium sulphate Na_2SO_4, ferric chloride $FeCl_3$?

Prove your answer by ionic equations of hydrolysis of the corresponding salts.

501. Derive molecular and ionic equations for hydrolysis of calcium acetate $Ca(CH_3COO)_2$ and sodium hypochlorite NaClO. Which salt is hydrolyzed stronger, if the ionization constant of CH_3COOH is 1.8×10^{-5} and that of HClO is 4×10^{-8}?

502. A solution of NaH_2PO_4 has a slightly acid, and a solution of Na_3PO_4 strongly alkaline reaction. Explain these facts and give the corresponding ionic equations.

503. When solutions of chromium trichloride $CrCl_3$ and sodium sulphide Na_2S are mixed, chromium hydroxide $Cr(OH)_3$ precipitates. Explain the formation of the precipitate and express the reaction by the ionic equation.

504. Derive ionic and molecular equations for the reaction between aluminium sulphate $Al_2(SO_4)_3$ and potassium sulphide K_2S, bearing in mind that aluminium sulphide Al_2S_3 hydrolyzes completely.

505. When concentrated solutions of ferric chloride $FeCl_3$ and soda Na_2CO_3 are mixed, carbon dioxide is liberated and $Fe(OH)_3$ precipitate is formed. Explain the reaction and express it by the ionic equation.

CHAPTER XIII

COMPLEX COMPOUNDS

Many positively charged ions can unite with opposite sign ions or polar molecules (NH_3, H_2O, etc.) to form the so-called *complex ions*. Compounds incorporating complex ions are called *complex compounds*. Complex salts are especially important.

The formation of a complex compound can be illustrated by the following examples.

1. If a solution of nickelous chloride $NiCl_2$ is reacted with ammonia, each Ni^{2+} ion joins in the solution with six molecules of ammonia to convert into the complex ion $[Ni(NH_3)_6]^{2+}$. As water is evaporated, the complex ions are bonded with chloride ions to produce the complex salt $[Ni(NH_3)_6]Cl_2$.

2. When a solution of potassium iodide KI acts on a mercury iodide HgI_2 precipitate, it is dissolved owing to the formation of the complex salt $K_2[HgI_4]$ which is present in the solution in the form of K^+ ions and $[HgI_4]^{2-}$ complex ions.

The elementary ion that combines with neutral molecules or ions of the opposite sign to form a complex ion is called the *complexing agent* or the *central ion*. In the above salts, the nickel and mercury ions are the complexing agents.

The complex compounds are characterized by a specific spatial arrangement of the component parts in a molecule. The central position in a molecule is occupied by the central ion, that is the complexing agent. Coordinated in the immediate vicinity are several other ions or neutral molecules, which form the so-called *inner coordination sphere* of the compound. The rest of the ions are farther away from the complexing agent and constitute the *outer coordination sphere*. During dissolution of a complex compound in water, the ions of the outer sphere are split off, whereas the ions or molecules coordinated in the inner sphere remain bonded with the central ion to form a stable, undissociating (or almost so) complex ion. In the formula of a complex compound, the complex ion is bracketed.

Ions or molecules bonded with the central ion are called *addends*. A complex ion can contain simultaneously various addends. For example, in the complex ion $[Co(NH_3)_2 (NO_2)_4]^-$ molecules of NH_3 and NO_2^- ions are the addends.

The charge on a complex ion equals the algebraic sum of the charges on its constituent simple ions. Neutral molecules do not influence the charge. For example, the charge of the ion $[Co(NH_3)_2(NO_2)_4]^-$ is summed up from the charges of Co^{3+} and NO_2^- ions and equals $3-4=-1$. The charge of the ion $[HgI_4]^{2-}$ is $2-4=-2$.

The structure of a complex compound is often represented by elaborated structural (coordination) formulas, such as

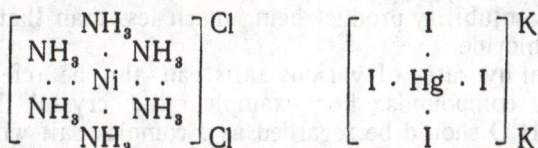

where the bonds between the addends and the central ion are marked by dots.

Each complexing agent can join only a definite number of addends. This number is called the *coordination number* of the complexing agent. For example, the coordination number of nickel is, as a rule, 6, that of mercury 4, of silver 2, etc.

Complex salts are very much like double salts and they are often formed by the combination of two simple salts. For example:

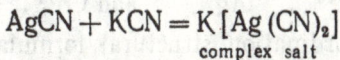
complex salt

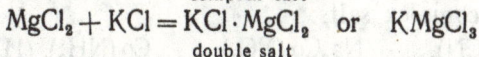
double salt

However, when dissolved in water, a double salt falls into the ions found in solutions of the constituent simple salts, whereas a complex salt dissociates to form new complex ions having their specific characteristics:

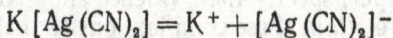
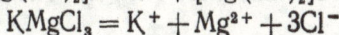

However, no sharp dividing line can be drawn between the two types of salts since many complex ions can dissociate to a greater or lesser degree with formation of simple ions. For example, in an aqueous solution, the complex salt $[Ag(NH_3)_2]NO_3$ dissociates into $[Ag(NH_3)_2]^+$ and NO_3^- ions. In turn, the $[Ag(NH_3)_2]^+$ ion dissociates (although to an insignificant degree) into a silver ion and molecules of ammonia:

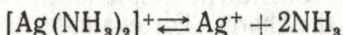

$$[Ag(NH_3)_2]^+ \rightleftarrows Ag^+ + 2NH_3$$

The number of silver ions in the solution is small and when common salt is added, it does not precipitate silver chloride, since the value of the solubility product of AgCl is not attained. But as soon as iodide ions are introduced into the solution, a yellow precipitate of silver iodide falls out immediately, its solubility product being much less than that in the silver chloride.

Crystal hydrates of various salts can also be referred to complex compounds. For example, the crystal hydrate $NiCl_2 \cdot 6H_2O$ should be regarded as a complex salt whose six molecules of water enter the composition of a complex ion: $[Ni(H_2O)_6]Cl_2$.

PROBLEMS

506. Determine the magnitude and the sign of the charge on the complex ions given below:

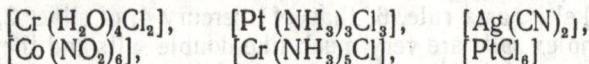

$[Cr(H_2O)_4Cl_2]$, $\qquad$ $[Pt(NH_3)_3Cl_3]$, $\qquad$ $[Ag(CN)_2]$,
$[Co(NO_2)_6]$, $\qquad$ $[Cr(NH_3)_5Cl]$, $\qquad$ $[PtCl_6]$

bearing in mind that the complexing agents here are ions

$$Cr^{3+}, \qquad Pt^{4+}, \qquad Ag^+, \qquad \text{and } Co^{3+}$$

507. Derive the coordination structural formulas of the following complex salts:

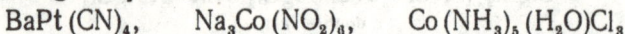

$$BaPt(CN)_4, \qquad Na_3Co(NO_2)_6, \qquad Co(NH_3)_5(H_2O)Cl_3$$

bearing in mind that the complexing agents in these salts are Co^{3+} and Pt^{2+} ions.

508. The coordination number of bivalent copper is 4. Derive the formulas for the ammonium and cyanide complexes of bivalent copper, specify their valence and give examples of salts that may contain these complex ions as their constituent parts.

509. Write equations for ionization of the following complex salts:

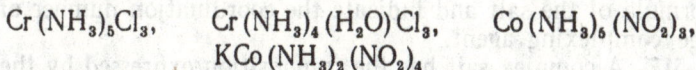

$$Cr(NH_3)_5Cl_3, \qquad Cr(NH_3)_4(H_2O)Cl_3, \qquad Co(NH_3)_5(NO_2)_3,$$
$$KCo(NH_3)_2(NO_2)_4$$

Place the complex ions in square brackets bearing in mind that the coordination number of both chromium and cobalt is 6.

510. What ions are complexing agents in the following complex salts:

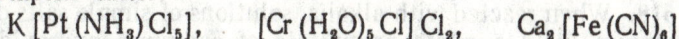

$$K[Pt(NH_3)Cl_5], \qquad [Cr(H_2O)_5Cl]Cl_2, \qquad Ca_2[Fe(CN)_6]$$

Determine their valence and the coordination numbers.

511. Silver nitrate precipitates all chlorine as AgCl from a solution of the complex salt $Pt(NH_3)_6Cl_4$ and only 1/4 part of chlorine from a solution of the salt $Pt(NH_3)_3Cl_4$. Write the formulas of the above salts placing the complex ions into square brackets and indicate the coordination number of platinum in these salts.

512. There are two complex salts of cobalt, both having the empirical formula $CoClSO_4 \cdot 5NH_3$. The difference between them is that a solution of one salt when reacting with barium chloride yields a precipitate of barium sulphate without precipitating silver nitrate, whereas the other salt, on the contrary, yields a precipitate of silver nitrate and does not precipitate sodium chloride. Write the coordination formulas of both salts and indicate the coordination number of cobalt.

513. Write molecular and ionic equations for the exchange reactions between $K_4Fe(CN)_6$ and $CuSO_4$, $Na_3[Co(CN)_6]$ and $FeSO_4$, $K_3[Fe(CN)_6]$ and $AgNO_3$, bearing in mind that the resultant complex salts are insoluble in water.

514. Write equations for ionization of the complex salts of cobalt placing the complex ions into square brackets:

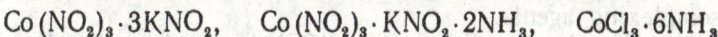

$$Co(NO_2)_3 \cdot 3KNO_2, \qquad Co(NO_2)_3 \cdot KNO_2 \cdot 2NH_3, \qquad CoCl_3 \cdot 6NH_3$$

515. Anhydrous chromium trichloride $CrCl_3$ combines with ammonia to form two complex salts, viz., $CrCl_3 \cdot 5NH_3$ and $CrCl_3 \cdot 6NH_3$. Write the coordination formulas and equations for ionization of these salts knowing that silver nitrate precipitates all chlorine from one of them, and only 2/3 chlorine from the other.

516. Silver nitrate precipitates 2/3 of all chlorine from a solution of the complex salt $CoCl_3 \cdot 5NH_3$. Write the structural formula of the salt and indicate the coordination number of the complexing agent.

517. A complex salt has the composition expressed by the formula $CoClSO_4 \cdot 5NH_3$. No precipitate is formed in the reaction between a solution of this salt and silver nitrate, while barium chloride precipitates barium sulphate from the solution. Write the equation for ionization of this salt placing the complex ion into square brackets.

518. When reacted with alkalis, solutions of simple salts of cadmium yield a white precipitate of cadmium hydroxide $Cd(OH)_2$, and with hydrogen sulphide they form a yellow precipitate of cadmium sulphide CdS. How can the fact be explained that while forming a precipitate with hydrogen sulphide a solution of cyanide complex of cadmium $K_2[Cd(CN)_4]$ does not produce precipitates with alkalis?

519. Potassium iodide KI precipitates silver as AgI from a solution of the complex salt $[Ag(NH_3)_2]NO_3$ and does not form any precipitate in a solution of the salt $K[Ag(CN)_2]$. At the same time both solutions, when reacting with hydrogen sulphide, yield a precipitate of silver sulphide Ag_2S. What conclusion can be drawn with respect to the dissociation of the complex ions $[Ag(NH_3)_2]^+$ and $[Ag(CN)_2]^-$? What salt has lesser solubility product, Ag_2S (silver sulphide) or AgI (silver iodide)?

520. What is the difference between complex and double salts from the standpoint of the ionic theory? Write equations for ionization of the complex and double salts given below:

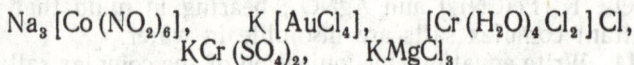

$$Na_3[Co(NO_2)_6], \qquad K[AuCl_4], \qquad [Cr(H_2O)_4Cl_2]Cl,$$
$$KCr(SO_4)_2, \qquad KMgCl_3$$

Indicate the valence and the coordination numbers of the complexing agents.

521. If potassium thiocyanate KSCN is added to a solution containing ferric ions, the solution is coloured intense scarlet due to the formation of ferric thiocyanate $Fe(SCN)_3$. Will the red colour develop if potassium thiocyanate is added to a solution of ferriammonium sulphate $NH_4Fe(SO_4)_2 \cdot 12H_2O$ or to a solution of potassium ferricyanide $K_3[Fe(CN)_6]$? Develop your reasons.

CHAPTER XIV

MENDELEYEV PERIODIC SYSTEM
OF ELEMENTS

1. Atom Structure and Properties of Elements

In accordance with the modern conceptions, the properties of chemical elements are dependent on the structure of their atoms. In the Mendeleyev Periodic System of Elements, the elements are arranged in order of their increasing charges on the nuclei of the atoms and divided into seven horizontal rows (three short and four long *periods*). The first period comprises two elements, the second and the third contain 8 elements each, the fourth and the fifth 18 elements, the sixth 32 elements and the seventh (incomplete) contains 18 elements. The atomic number of the element indicates the charge on the atom nucleus and, at the same time, the number of electrons revolving around it. Electrons form several shells. The maximum number of electrons in each shell (N) is determined by the formula $N = 2n^2$, where n is the number of the shell.

As the nuclear charge increases, both the total number of electrons in the atom and the number of electron shells increase too. How new electron shells are constructed and filled with electrons, can be seen in the Table of the Periodic System which is appended to this textbook. While examining the Table one can notice that a new electron shell appears in each period, and in long periods the *preceding* shells are completed with electrons in addition. Therefore, *the total number of electron shells in the atom equals the number of the period in which the element stands*.

The properties of elements change regularly with their increasing atomic numbers. The most specific chemical property inherent in metals is the ability of their atoms to give off electrons from their outer shell and to convert into positively charged ions. Nonmetals, on the contrary, are characterized by their ability to accept electrons and thus form negatively charged ions.

Let us consider the periodicity of changes in the properties of elements inside a period.

Within the boundaries of a period (except the first one) the metallic character, which is the most pronounced in the first element, decreases gradually towards the end of the period, whereas the nonmetallic nature increases. Therefore, each period begins with a typical metal and a typical nonmetal stands in the end of the period, after which follows a noble gas.

The regular character of changes in properties of the elements in periods is explained by that with growing charge on the atom nucleus, the force that attracts the electrons toward it increases, and the property to donate electrons, which is characteristic of metals, decreases. At the same time, atoms begin to show the tendency to taking up electrons, which becomes more apparent in the end of the period.

Apart from the horizontal division, all elements in the Periodic Table are also divided into nine vertical series (*groups*). The first seven groups are subdivided into the main and the secondary subgroups. The number of electron shells in atoms increases inside a subgroup with the atomic number of the element and since the electrons in the outer (valence) shells are located farther and farther from the nucleus, their attraction to it diminishes and the electrons can be easily detached from the atom. Therefore, with increasing atomic number of the element inside a subgroup, the metal properties as a rule grow stronger and the nonmetal properties weaken.

These regularities allow one to determine the position of an element in the Periodic System from its atomic number, as well as to make out the structure of the electronic shells of its atom and to outline the main chemical properties of the element.

Example 1. What is the structure of electron shells in an element having the atomic number 33?

Solution. Consider first the position of the element in the Periodic Table.

This element is located in the fourth period, the odd series, the fifth group. Since the third electron shell in the even series (in its first half) of the fourth period is completed to 18 electrons, the first three electron shells in all elements of the odd series of the fourth period are completed, the number of electrons in the fourth shell is equal to the number of the group. It follows therefore that the structure of the electron shells in the element in question will be as this: the first shell contains 2 electrons, the second shell 8 electrons, the third 18 electrons and the fourth 5 electrons.

Example 2. Describe the main chemical properties of an element, whose atomic number is 52, from the structure of its atom.

Solution. The presence of six electrons in the outer shell of the atom indicates that nonmetal properties must predominate in this element. However, owing to the great distance of the outermost electrons from the nucleus the nonmetal properties should be somewhat weakened.

The highest positive valence of the element should be equal to 6, and the negative to 2. The formula of the highest oxide should be RO_3. The element should form a gaseous compound of the RH_2 type with hydrogen.

PROBLEMS

522. Which elements are characterized by the formation of gaseous compounds with hydrogen? In which groups of the Periodic Table should these elements reside? Indicate, which of their compounds with hydrogen possess acid properties.

523. The element whose highest salt-forming oxide has the formula R_2O_5 forms a gaseous compound containing 3.85 per cent of hydrogen. What is this element?

524. When 0.75 g of a bivalent metal reacts with water, 420 ml of hydrogen are liberated (as measured at STP). Name the metal.

525. Draw the electron diagrams of atomic structure of the first four elements in the fourth period, indicate the highest valence of these elements and designate their ions.

526. Draw electron diagrams of atomic structure of elements having the atomic numbers 14, 15, 16 and 17. Indicate the highest possible positive and negative valence of each element.

527. Draw electron diagrams of atomic structure of negative ions of sulphur and chlorine. Although the structure of the electron shells in these ions is identical, they have quite different properties. Why?

528. Manganese, the element of the seventh group, has predominantly metal properties, whereas the halogens standing in the same group are typical nonmetals. Explain this phenomenon from the viewpoint of atomic structure of these elements.

529. To what element, selenium or chromium, does molybdenum bear a greater resemblance with respect to its atomic structure? What properties, metal or nonmetal, should dominate in it? Explain your answer.

530. In what element, calcium or barium, are the metal properties stronger? Why? Which of them forms a stronger base?

531. How do the properties of elements change in periods and in the main subgroups of the Periodic Table with increasing atomic number? Explain these changes from the aspect of the atomic structure of these elements.

532. Which element of the fourth group, titanium or germanium, has stronger metal properties? Prove your answer from the aspect of the structure of their electron shells.

533. Elements have 2 electrons in the outermost shell and 13 in the shell next to the outermost one. In what group and in what series, odd or even, do these elements stand? What is their maximum possible valence? What properties, metallic or nonmetallic, dominate in them?

534. What properties does an imaginary element possess if its atomic number is 87? To what known element in the Table can it be compared?

535. The maximum valence of an element with respect to oxygen is 7. The outer electron shell of its atom contains two electrons. Does this element form a gaseous compound with hydrogen?

536. How can the periodic change in the valence of elements with their increasing atomic numbers be explained? Why does the valence remain unaltered in the elements Nos. 58 through 71 (lanthanides) with their increasing atomic numbers?

537. Indicate ions of elements in the third period comparable to neon with respect to the structure of their electron shells. Designate them with the appropriate symbols.

538. What ions of the first four elements in the fourth period are similar with respect to their atomic structure to an argon atom? Designate them with the appropriate symbols.

539. Indicate ions in which the structure of the outermost shells is similar: Cd^{2+} and Ag^+ or Cd^{2+} and Ba^{2+}.

540. How many electrons do Ca^{2+}, Cr^{3+}, Zn^{2+}, Se^{2-} and Br^- ions of the fourth period elements contain in their outer shells? Which of them are similar with respect to the structure of the electron shell to the atoms of the inert gases?

541. How many electrons do the ions

$$Rb^+, \quad Sr^{2+}, \quad Zr^{4+}, \quad Ag^+, \quad Cd^{2+}, \quad Sn^{4+} \text{ and } Te^{2-}$$

of the elements standing in the fifth period contain in their outermost shells? Which of these ions resemble the atom of the noble gas krypton with respect to the structure of their electron shells?

542. How many electron shells do K^+, Mn^{2+}, Fe^{3+}, I^- and Se^{2-} ions contain?

543. What is the maximum valence with respect to oxygen and the valence with respect to hydrogen of elements having in their two outer shells 18 and 5 (one element) and 8 and 7 electrons (the other element)?

544. Draw schemes of electron shells of the following ions:

$$Li^+, \quad Cs^+, \quad Ba^{2+}, \quad Al^{3+}, \quad Ni^{2+}, \quad Cl^-, \quad S^{2-}$$

545. Is the structure of electron shells identical (a) in chloride and calcium ions in crystals of calcium chloride? (b) in bromide and sodium ions in crystals of sodium bromide?

2. Radioactivity.
Displacement Law. Isotopes

Radioactivity is the property of certain elements to emit rays that can penetrate various substances, otherwise impermeable to ordinary light, to ionize air and darken photographic plates wrapped in black paper.

Three types of rays emitted by radioactive elements are distinguished:

(1) α-rays; these are streams of particles whose mass is four carbon units, each particle carrying two elementary positive charges (nuclei of helium);

(2) β-rays; these are streams of electrons having the velocity approximating that of light;

(3) γ-rays; these are electromagnetic oscillations of a very high frequency; they are similar to Roentgen rays, but have a much shorter wavelength.

Emission of rays by the radioactive elements is the result of spontaneous decay of nuclei of their atoms into nuclei of atoms of new chemical elements. The decay of atoms is accompanied by emission of alpha particles (nuclei of helium) and beta particles (electrons).

The quantity of atoms of a radioactive element, breaking up at each given moment, is proportional to the quantity of atoms actually present (**The law of radioactive decay**).

The proportionality coefficient, showing the fraction of the total number of atoms of the radioactive element actually present that disintegrate per unit time, is called the *radioactive constant* of an element. Knowing the radioactive constant, one can calculate the number of disintegrated atoms at any given moment of time. For example, if 1/100 part of the radioactive element disintegrates per second, then 1/100 part of the remaining element will disintegrate next second, and during the next second 1/100 part of the new residue will disintegrate, and so on.

The time necessary for half the initial quantity of a radioactive element to decay is called its *half-life period*.

The value which is reverse to the radioactive constant is called the *mean lifetime* of a radioactive element. It is evident that this is the period of time during which any quantity of a radioactive element will disintegrate without residue if the decay proceeds at the same (initial) rate. The mean lifetime (τ) relates to the half-life period (T) as this: $\tau = 1.44T$.

Arrangement of the elements that are formed in radioactive disintegration in the Periodic Table and also determination of their atomic weights are guided by the so-called *displacement law*, the essence of which is as follows. As a nucleus of an atom emits one alpha particle, the mass of the atom decreases 4 units and the charge of the nucleus 2 units. The atomic weight of the newly formed element is therefore 4 units less and the atomic number 2 units less than the parent element, and the new element is shifted in the Periodic Table two places to the left of the parent element. Emission of a beta particle from a nucleus does not change the atomic weight since the mass of a beta particle is negligibly small (1/1840 mass of the hydrogen atom), but the charge on the nucleus increases one unit; as a result, the newly formed element is displaced in the Periodic Table one place to the right of the parent element.

Example 1. Determine the atomic weight and the place occupied in the Periodic Table by the element which is formed on disintegration of radium during which four alpha and two beta particles were emitted.

Solution. The atomic number of radium is 88. The emission of four alpha particles reduces the charge on the nucleus 8 units, and the emission of two beta particles increases it 2 units. It follows therefore that the atomic number of the newly formed element should be 6 units less than that of radium, that is 82.

Example 2. An intermediate product of thorium decay is thoron whose atomic weight is 220 and the atomic number 86. How many alpha and beta particles are emitted as an atom of thorium is converted into an atom of thoron?

Solution. The atomic weight of thorium is 232. The difference between the atomic weights of thorium and thoron is 12. Hence, during the conversion of thorium into thoron, the former should emit three alpha particles.

The atomic number of thorium is 90. If the conversion of thorium into thoron were accompanied with emission of only alpha particles, the atomic number of thoron would be 84. But it is 86; hence two beta particles must be emitted in addition to the three alpha particles.

* * *

Atoms having identical charges and hence identical chemical properties, but different atomic weights, are called *isotopes*.

The majority of chemical elements are mixtures of isotopes, in other words, they contain atoms having different mass. Therefore, the atomic weight of an element, as determined by the usual chemical methods, is only an arithmetic mean of atomic weights of its isotopes. Atomic weights of isotopes are expressed by almost accurately. whole numbers.

The isotopes of one and the same element are denoted by the usual symbols used for the corresponding element with the addition of a superscript index in their upper right corner indicating the atomic weight of the isotope. For example, the isotopes of chlorine are denoted by Cl^{35} and Cl^{37}.

PROBLEMS

546. During one hour, 1/6 part of a radioactive element is disintegrated. Make out a graph, by plotting the reduction of quantity of this element against time. Determine the half-life period of this element using the graph.

547. The radioactive constant of radium E (bismuth isotope) is 1.6×10^{-6} (per second). Determine the mean lifetime of this element.

548. The radioactive constant of radium A (polonium isotope) is 3.8×10^{-3} (per second). Determine the mean lifetime and the half-life of this element.

549. A vessel contains 0.01 g of radon. The half-life period of radon is approximately 4 days. What quantity of radon will remain in the vessel in 20 days?

550. One gram of radium emits 3.5×10^{10} alpha particles per second. Calculate, how long it will take for this amount of radium to decompose completely, if the disintegration rate remains constant. How is this period called?

551. There is 1 mg of radium C (bismuth isotope), the half-life period of which is approximately 20 minutes. What quantity of radium C will remain in three hours?

552. How can it be explained that although continually disintegrating, radium still occurs in nature?

553. The half-life period of radium is 1580 years. In how many years will one gram of radium be reduced to 1/8 g?

554. In progressive radioactive conversions of a nucleus, one alpha and two beta particles were formed. Will the mass and the charge of the nucleus change as a result of these conversions?

555. How many alpha and beta particles will be formed in the conversion of an atom of lead Pb^{208} into an atom of mercury Hg^{200}?

556. Potassium and rubidium are slightly radioactive and emit beta rays. Into what elements are they converted?

557. Into what group of the Periodic Table should the element produced by progressive disintegration of thorium be placed if four alpha and two beta particles were emitted? What are its atomic weight, valence, the number of electrons in the outermost shell, the formula of the highest oxide and the formula of its compound with hydrogen?

558. To what group of the Periodic System does the element produced from uranium U^{238} belong, if in the progressive disintegration of the uranium, two alpha particles and two beta particles were emitted? What is the atomic weight of this element? What element of the Periodic Table has similar chemical properties?

559. What volume will helium formed due to conversion of one gram-atom of thorium into lead Pb^{208} occupy at STP? The atomic weight of thorium is 232. It should be remembered that a helium molecule is monoatomic.

560. How many grams of lead Pb^{206} and how many litres of helium at STP will be obtained in disintegration of a gram-atom of radium?

561. How can atoms of one and the same element differ? How do we call these atoms? Can atoms of different elements have equal weights? Exemplify your answers.

562. Nitrogen consists of isotopes N^{14} and N^{15}, and hydrogen of isotopes H^1 and H^2. How many different types of molecules does ammonia contain? Write their formulas (for example, $N^{14}H^2H_2^1$) and indicate the molecular weights.

563. Oxygen consists of isotopes O^{16}, O^{17} and O^{18}, and carbon of isotopes C^{12} and C^{13}. How many types of different molecules does a carbon dioxide molecule contain? Write their formulas and indicate the molecular weights.

564. An isotope of what element will be obtained in the progressive disintegration of an atom of thorium Th^{232} during which five alpha and two beta particles are emitted? What will the atomic weight of this isotope be?

565. Chlorine (atomic weight 34.46) consists of isotopes Cl^{35} and Cl^{37}. What are the percent concentrations of the isotopes?

566. Naturally occurring gallium (atomic weight 69.72) consists of isotopes Ga^{69} and Ga^{71}. What is the percent concentrations of the isotopes in gallium?

567. Naturally occurring rubidium consists of 72.8 per cent of Rb^{85} and 27.2 per cent of Rb^{87} isotopes. Calculate the mean atomic weight of rubidium.

568. Naturally occurring copper consists of isotopes Cu^{63} and Cu^{65} contained in the ratio of 8 : 3. Calculate the mean atomic weight of copper.

569. Actinium (atomic weight 227) is converted into actinon according to this scheme:

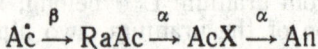

$$Ac \xrightarrow{\beta} RaAc \xrightarrow{\alpha} AcX \xrightarrow{\alpha} An$$

Indicate the atomic weight and atomic number for each intermediate member in the series. What elements in the Periodic Table are the isotopes of these intermediates?

570. The radioactive series uranium-radium is as this:

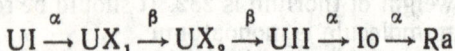

$$UI \xrightarrow{\alpha} UX_1 \xrightarrow{\beta} UX_2 \xrightarrow{\beta} UII \xrightarrow{\alpha} Io \xrightarrow{\alpha} Ra$$

Calculate the atomic weight and the atomic number of each intermediate member in the series. What elements in the Periodic Table are the isotopes of these intermediates?

CHAPTER XV

OXIDATION-REDUCTION (REDOX) REACTIONS

1. Oxidation and Reduction

Reactions during which the valence of elements of the reacting substances changes are called oxidation-reduction, or redox, reactions. They are accompanied by a transfer of electrons from one set of atoms or ions to another owing to which the valence is changed.

Oxidation is the process in which atoms or ions lose their electrons, and during reduction they gain them.

The substances whose atoms or ions gain electrons during the reaction are called *oxidizing agents* or *oxidants*, while the substances that donate electrons are *reducing agents* or *reductants*. During the reaction the oxidizing agent abstracts electrons from the substance oxidized and is itself reduced, and conversely the reducing agent loses electrons and is thus oxidized. Since the electrons given off by one substance are immediately accepted by another, the oxidation is always accompanied by the reduction, and vice versa.

The simplest oxidation-reduction reactions are the formation of a complex substance out of simple ones, and also the displacement of elements from their compounds by other elements.

Let us consider a few examples.

1. *Formation of zinc sulphide* ZnS *from sulphur and zinc* is expressed by the equation:

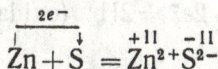

$$\overset{2e^-}{\overbrace{Zn + S}} = \overset{+11}{Zn^2} + \overset{-11}{S^{2-}}$$

The essence of the reaction is that two electrons of a zinc atom transfer to an atom of sulphur, owing to which both atoms turn into ions which form zinc sulphide. Zinc is oxidized, its valence increasing from 0 to $+2$. Sulphur is reduced,

its valence decreases from 0 to —2. The process can be represented graphically by expressing the oxidation and reduction
by separate "electronic" equations:

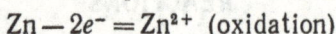

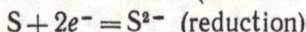

$$Zn - 2e^- = Zn^{2+} \text{ (oxidation)}$$
$$S + 2e^- = S^{2-} \text{ (reduction)}$$

Zinc is the reducing agent and sulphur is the oxidant.

2. *Displacement of copper from its salt* ($CuSO_4$) *with iron*:

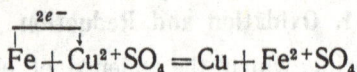

$$\overset{2e^-}{\overbrace{}}$$
$$Fe + Cu^{2+}SO_4 = Cu + Fe^{2+}SO_4$$

Copper ion Cu^{2+} takes two electrons from the iron atom and
turns into an electrically neutral atom of copper, whereas the
atom of iron turns into a positive doubly charged ion:

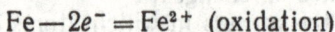

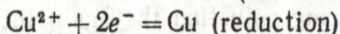

$$Fe - 2e^- = Fe^{2+} \text{ (oxidation)}$$
$$Cu^{2+} + 2e^- = Cu \text{ (reduction)}$$

Iron is oxidized, its valence grows from 0 to +2; copper ion
(Cu^{2+}) is reduced, its valence decreases from + 2 to 0. The
iron is the reducing agent and the copper ion is the oxidizer.

3. *Reduction of copper oxide with hydrogen*:

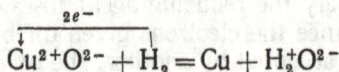

$$\overset{2e^-}{\overbrace{}}$$
$$Cu^{2+}O^{2-} + H_2 = Cu + H_2^+O^{2-}$$

In this reaction the copper ion Cu^{2+} captures two electrons
from two hydrogen atoms, which form the molecule H_2, and
converts into a neutral atom of copper, whereas hydrogen atoms
lose their electrons and convert into positive singly charged
ions of hydrogen H^+. The copper ion is the oxidizing agent and
hydrogen the reducing agent.

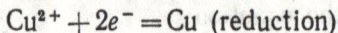

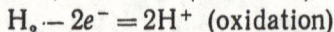

$$Cu^{2+} + 2e^- = Cu \text{ (reduction)}$$
$$H_2 - 2e^- = 2H^+ \text{ (oxidation)}$$

2. Equations of Oxidation-Reduction Reactions

Since in the redox reactions electrons are only transferred
from one set of atoms to another, it is evident that the number
of electrons captured by the oxidizer is equal to that given

off by the reducing agent. Therefore, while making out equations for reactions between oxidizing and reducing substances, the coefficients in the formulas should be so selected that the product of the number of the reacting atoms or molecules of the oxidant and the number of the captured electrons should be equal to the analogous product for the reducing substance. The number of captured or lost electrons is determined by the change in the valence of the corresponding elements.

While deriving equations for oxidation-reduction reactions one should know into what products the oxidizing and reducing agents will be converted. Very often it can be easily predicted from the properties of both substances. If the reaction proceeds in solution, water can also take part in the reaction, it becoming clear in making out the equation.

Let us consider a few examples of deriving equations for oxidation-reduction reactions:

1. *Oxidation of hydrogen iodide with concentrated sulphuric acid.* The reaction proceeds according to the scheme:

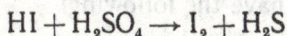

$$HI + H_2SO_4 \rightarrow I_2 + H_2S$$

Determine the valence of the elements participating in the oxidation and reduction processes before and after the reaction and denote it by the Roman numerals over the symbols of the elements:

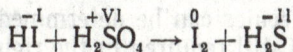

$$\overset{-I}{H}I + H_2\overset{+VI}{S}O_4 \rightarrow \overset{0}{I_2} + H_2\overset{-II}{S}$$

The valence of iodine increases in the reaction from -1 to 0; hence the iodine (iodide ion) is oxidized and each of its ions loses one electron. The valence of sulphur is reduced from $+6$ to -2, that is sulphur is reduced in the reaction. The valence decreases due to the transfer of electrons from iodide ions to sulphuric acid. It can be assumed that each atom of sulphur having covalent bonds with oxygen atoms in the SO_4^{2-} ion gained eight electrons to turn into S^{2-} ion which is part of the hydrogen sulphide molecule. Let us designate this atom as $\overset{+VI}{S}$ and describe the transfer of electrons in the form of the following electronic equations:*

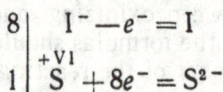

$$8 \left| \begin{array}{l} I^- - e^- = I \\ \overset{+VI}{S} + 8e^- = S^{2-} \end{array} \right.$$

The number of electrons given off by the iodide ions must be equal to that of electrons gained by sulphur atoms $\overset{+VI}{S}$, therefore each atom $\overset{+VI}{S}$ requires eight iodide ions, in other words one molecule of sulphuric acid will react with eight molecules of hydrogen iodide. Hence, the corresponding numerals should be placed to the left of the equation.

Thus we find the coefficients for the oxidizing agent (H_2SO_4) and the reducing agent (HI), and also the coefficients for substances obtained in the reaction, since it is quite apparent that four molecules of iodine will be produced out of eight molecules of hydrogen iodide, and only one molecule of hydrogen sulphide from one molecule of sulphuric acid. By placing the found coefficients into the unbalanced scheme of the reaction we have the following:

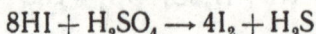

$$8HI + H_2SO_4 \longrightarrow 4I_2 + H_2S$$

If we now compare the number of hydrogen atoms in the left-hand and the right-hand parts of the scheme, we shall find that eight hydrogen atoms more are still required and oxygen is absent in the right-hand part of the scheme. It is evident that water is also produced in the reaction and the number of its molecules can be determined from the number of hydrogen atoms still required in the right-hand part of the scheme. If we now place four molecules of water in the right-hand part, the equation will be balanced and the arrow can now be replaced by the sign of equality:

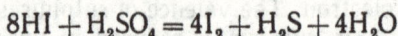

$$8HI + H_2SO_4 = 4I_2 + H_2S + 4H_2O$$

The correctness of the equation can be checked by putting together the number of oxygen atoms in its both parts.

---

* Atoms having covalent bonds in molecules of complex substances or in complex ions will be further designated by the symbols of the corresponding elements with a Roman numeral above them which will indicate the valence of the element (atom).

2. *Reduction of ferric chloride* $FeCl_3$ *with stannous chloride* $SnCl_2$. The unbalanced scheme of this reaction is as this:

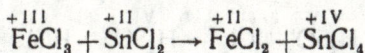

$$\overset{+III}{FeCl_3} + \overset{+II}{SnCl_2} \longrightarrow \overset{+II}{FeCl_2} + \overset{+IV}{SnCl_4}$$

It follows from the reaction scheme that the valence of iron decreases from $+3$ to $+2$, and the valence of tin increases from $+2$ to $+4$. Hence, the ferric iron (Fe^{3+}) is reduced and tin (Sn^{2+}) is oxidized in the reaction due to the electron transfer from tin ions to ferric ions, during which each tin ion Sn^{2+} loses two electrons and each ferric ion Fe^{3+} gains one electron. By representing these processes in the form of electronic equations we can derive (as in the previous example) a scheme illustrating the redistribution of electrons and find the coefficients for the equation:

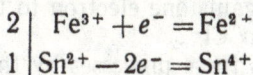

$$\begin{array}{c|l} 2 & Fe^{3+} + e^- = Fe^{2+} \\ 1 & Sn^{2+} - 2e^- = Sn^{4+} \end{array}$$

Numbers 2 and 1 are the coefficients in the formulas of substances in the left-hand part of the equation. Taking them into consideration we can now calculate the coefficients for the right-hand part of the equation and replace the arrow by the equality sign:

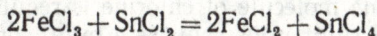

$$2FeCl_3 + SnCl_2 = 2FeCl_2 + SnCl_4$$

The equal number of identical atoms in the right-hand and the left-hand sides of the equation shows that the equation is balanced correctly.

Since the reaction in question proceeds between aqueous solutions of electrolytes, it can also be expressed by the ionic equation:

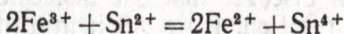

$$2Fe^{3+} + Sn^{2+} = 2Fe^{2+} + Sn^{4+}$$

When checking correctness of the ionic equations, it should be remembered that in any ionic equation, not only the total number of atoms or ions of each element in both parts of the equation but also the algebraic sums of charges in both parts of the equation should be equal. For example, in our case the equation has been derived correctly since $3 \times 2 + 2 = 2 \times 2 + 4$.

3. *Oxidation of sulphurous acid with chlorine*. Schematically this reaction can be represented as this:

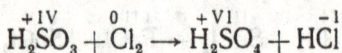

$$\overset{+IV}{H_2SO_3} + \overset{0}{Cl_2} \longrightarrow \overset{+VI}{H_2SO_4} + \overset{-1}{HCl}$$

In the reaction the valence of sulphur increases from $+4$ to $+6$, and that of chlorine decreases from 0 to -1.

The valence of sulphur increases owing to capturing by a chlorine molecule of two electrons donated by a sulphuric acid molecule. It may be assumed that the two electrons are donated by the atom of sulphur $\overset{+IV}{S}$ (which has a covalent bond with the atoms of oxygen in a molecule of sulphurous acid) and it is thus converted into a hexavalent atom of sulphur $\overset{+VI}{S}$ to enter the molecule of sulphuric acid. Simultaneously, each chlorine atom gains one electron to turn into negatively charged chloride ion Cl^-.

If we now express these processes by the electronic equations, we can find the coefficients to the formulas of the oxidizing and the reducing agents

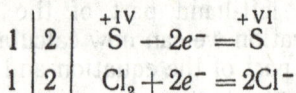

$$
1 \,\bigg|\, 2 \,\bigg|\,
\begin{array}{l}
\overset{+IV}{S} - 2e^- = \overset{+VI}{S} \\
Cl_2 + 2e^- = 2Cl^-
\end{array}
$$

showing that one molecule of chlorine is required to oxidize one molecule of sulphurous acid; as a result, one molecule of sulphuric acid and two molecules of hydrogen chloride are produced. By adding the coefficient 2 to HCl we get the following:

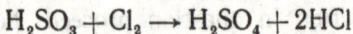

$$H_2SO_3 + Cl_2 \longrightarrow H_2SO_4 + 2HCl$$

If we count now the hydrogen atoms in both parts of the scheme, we shall discover that one molecule of water must take part in the reaction. If we put it into the left-hand part of the scheme, it will turn into a balanced equation:

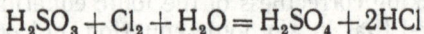

$$H_2SO_3 + Cl_2 + H_2O = H_2SO_4 + 2HCl$$

* * *

An acid can very often serve as an oxidizing agent in the oxidation-reduction reactions. Apart from the oxidizing function, its molecules can also bond the ions which are formed in the reaction. For example, during oxidation of metals, the molecules of an acid bond ions of the metal to form a salt. In deriving the final equation for such reactions, the corresponding number of molecules spent for the formation of the salt should be added to the number of acid molecules found from the electronic equation.

Let us consider the following reaction.

4. *Oxidation of magnesium with dilute nitric acid.* The products of the oxidation-reduction reaction between magnesium and dilute nitric acid are magnesium nitrate $Mg(NO_3)_2$ and nitrous oxide N_2O. The reaction can be schematically represented as this:

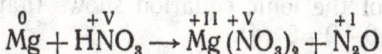

$$\overset{0}{Mg} + \overset{+V}{H}\overset{}{N}O_3 \rightarrow \overset{+II}{Mg}(\overset{+V}{N}O_3)_2 + \overset{+1}{N_2}O$$

As appears from the scheme, magnesium is oxidized (its valence increases from 0 to $+2$) and nitrogen is reduced (its valence decreases from $+5$ to $+1$).

By making out the electronic equations for the processes of oxidation and reduction

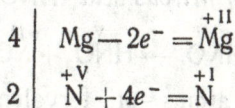

$$
\begin{array}{c|c}
4 & Mg - 2e^- = \overset{+II}{Mg} \\
2 & \overset{+V}{N} + 4e^- = \overset{+1}{N}
\end{array}
$$

we can find that in order to oxidize four atoms of magnesium two molecules of nitric acid are required which produce one molecule of nitrous oxide on the interaction with magnesium. At the same time, four molecules of magnesium nitrate must be formed which require another 8 molecules of nitric acid. Thus, the total number of nitric acid molecules that will participate in the reaction is ten. Of this number only two will be spent to oxidize magnesium, since the valence is decreased only in two nitrogen atoms (from $+5$ to $+1$), while the other eight molecules of the acid will be spent in the formation of the salt $Mg(NO_3)_2$. After introduction of the found coefficients the scheme of the reaction turns into

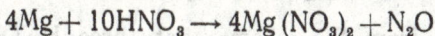

$$4Mg + 10HNO_3 \rightarrow 4Mg(NO_3)_2 + N_2O$$

Now the left-hand part of the scheme has ten hydrogen atoms,

whereas in the right-hand part hydrogen is absent. It follows therefore that five molecules of water are formed in the reaction. If we add the missing five molecules of water into the right-hand part of the scheme we balance it finally:

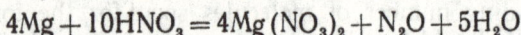

$$4Mg + 10HNO_3 = 4Mg(NO_3)_2 + N_2O + 5H_2O$$

The ionic equation for this reaction is:

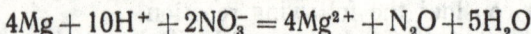

$$4Mg + 10H^+ + 2NO_3^- = 4Mg^{2+} + N_2O + 5H_2O$$

It becomes now quite apparent that only two NO_3^- ions, that is two molecules of nitric acid, are spent to oxidize four atoms of magnesium. Note that in the molecular equation this number is ten. Calculation of positive and negative charges in both parts of the ionic equation shows that the equation is correct $(+10—2=2\times4)$.

* * *

A specific case of the oxidation-reduction process is when the oxidizing agent and the reductant are atoms or ions of one and the same element. This process is known as the auto-oxidation-reduction.

5. *Decomposition of nitrous acid* HNO_2 *with heating*:

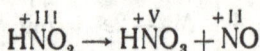

$$\overset{+III}{H}NO_2 \rightarrow \overset{+V}{H}NO_3 + \overset{+II}{N}O$$

In this reaction, atoms of tervalent nitrogen $\overset{+III}{N}$ are both oxidizing and reducing agents, in other words, a part of nitrous acid molecules play the role of the oxidizer and the other part that of the reducing agent. In the course of the reaction the former molecules are reduced and converted into NO molecules, whereas the latter are oxidized to HNO_3 molecules.

Now we can derive an ordinary electronic scheme and find the coefficients for the formulas of the oxidizer and the reducing agent, which are summed up in this case:

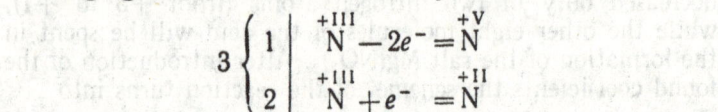

$$3 \begin{cases} 1 & \overset{+III}{N} - 2e^- = \overset{+V}{N} \\ 2 & \overset{+III}{N} + e^- = \overset{+II}{N} \end{cases}$$

It follows from the scheme that three HNO_2 molecules should be taken to balance the chemical equation. Of this quantity, one molecule of nitric acid and two molecules of nitric oxide are produced:

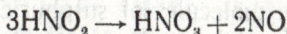

$$3HNO_2 \longrightarrow HNO_3 + 2NO$$

By comparing the number of hydrogen atoms in both parts of the equation we find that one molecule of water must be formed in the reaction. Hence the balanced equation of the reaction will be

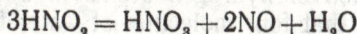

$$3HNO_2 = HNO_3 + 2NO + H_2O$$

* * *

The most complicated are oxidation-reduction reactions in which atoms or ions of two or several elements are oxidized or reduced simultaneously. The following reaction illustrates this particular case.

6. *Oxidation of arsenic trisulphide* As_2S_3 *with nitric acid*:

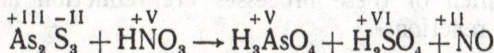

$$\overset{+III}{As_2}\overset{-II}{S_3} + \overset{+V}{HNO_3} \longrightarrow \overset{+V}{H_3AsO_4} + \overset{+VI}{H_2SO_4} + \overset{+II}{NO}$$

Arsenic and sulphur that compose the arsenic trisulphide are oxidized simultaneously in this reaction. The valence of arsenic increases from $+3$ to $+5$, and the valence of sulphur increases from -2 to $+6$. Nitrogen of nitric acid $(\overset{+V}{N})$ whose valence decreases from $+5$ to $+2$ (in the nitric oxide) is the oxidizing agent.

The electrons are redistributed as this:

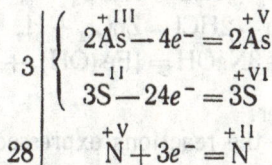

$$3 \begin{cases} \overset{+III}{2As} - 4e^- = \overset{+V}{2As} \\ \overset{-II}{3S} - 24e^- = \overset{+VI}{3S} \end{cases}$$

$$28 \left| \quad \overset{+V}{N} + 3e^- = \overset{+II}{N} \right.$$

Each molecule of As_2S_3, that is two atoms of $\overset{+III}{As}$ and three atoms of $\overset{-II}{S}$, donates 28 electrons*, and one molecule of HNO_3 accepts three electrons. Hence, twenty-eight mole-

---

* In order to find the correct coefficients in making out the electronic equations, the number of atoms, or ions, of each element must be equal to that contained in a molecule of the oxidizing or reducing substance.

cules of nitric acid are spent to oxidize three molecules of arsenic trisulphide. It is evident that six molecules of arsenic acid H_3AsO_4, nine molecules of sulphuric acid and twenty-eight molecules of nitric oxide are formed in the reaction. After placing thus found coefficients we get

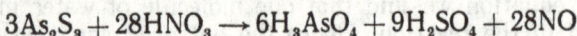

$$3As_2S_3 + 28HNO_3 \longrightarrow 6H_3AsO_4 + 9H_2SO_4 + 28NO$$

The calculation of hydrogen atoms in both parts of the equation shows that four molecules of water take part in the reaction. They should be included into the left-hand part of the equation to balance it:

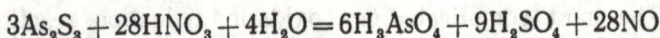

$$3As_2S_3 + 28HNO_3 + 4H_2O = 6H_3AsO_4 + 9H_2SO_4 + 28NO$$

PROBLEMS

571. Which of these processes are reduction and which oxidation reactions?

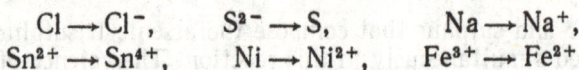

$$Cl \longrightarrow Cl^-, \qquad S^{2-} \longrightarrow S, \qquad Na \longrightarrow Na^+,$$
$$Sn^{2+} \longrightarrow Sn^{4+}, \qquad Ni \longrightarrow Ni^{2+}, \qquad Fe^{3+} \longrightarrow Fe^{2+}$$

572. Which reactions out of those expressed by the equations given below are oxidation-reduction reactions?

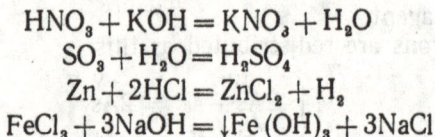

$$HNO_3 + KOH = KNO_3 + H_2O$$
$$SO_3 + H_2O = H_2SO_4$$
$$Zn + 2HCl = ZnCl_2 + H_2$$
$$FeCl_3 + 3NaOH = \downarrow Fe(OH)_3 + 3NaCl$$

Justify your answer.

573. Indicate for the reactions expressed by the equations

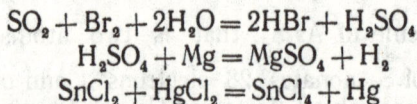

$$SO_2 + Br_2 + 2H_2O = 2HBr + H_2SO_4$$
$$H_2SO_4 + Mg = MgSO_4 + H_2$$
$$SnCl_2 + HgCl_2 = SnCl_4 + Hg$$

(a) what elements change their valence in the reaction; (b) what substances (and which particular elements) are oxidizing and which are reducing agents.

574. Specify the electron transfer for each of the following oxidation-reduction reactions:

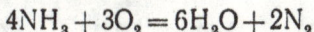

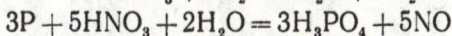

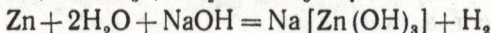

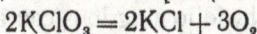

$$4NH_3 + 3O_2 = 6H_2O + 2N_2$$
$$3P + 5HNO_3 + 2H_2O = 3H_3PO_4 + 5NO$$
$$Zn + 2H_2O + NaOH = Na\,[Zn\,(OH)_3] + H_2$$
$$2KClO_3 = 2KCl + 3O_2$$

Which substances (elements) are oxidizing and which are reducing agents?

575. Derive electronic equations for the processes of oxidation and reduction for each of the following reactions:

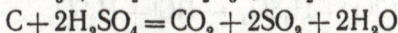

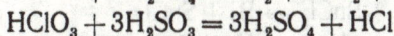

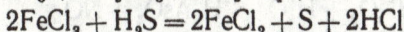

$$2AsH_3 + 3O_2 = As_2O_3 + 3H_2O$$
$$C + 2H_2SO_4 = CO_2 + 2SO_2 + 2H_2O$$
$$HClO_3 + 3H_2SO_3 = 3H_2SO_4 + HCl$$
$$2FeCl_3 + H_2S = 2FeCl_2 + S + 2HCl$$

Which of these substances (elements) are oxidizing and which are reducing agents?

576. Derive electronic equations and describe the role played by the atoms and ions of hydrogen in the following redox reactions:

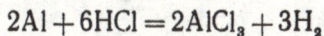

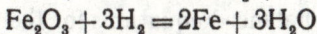

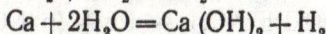

$$2Al + 6HCl = 2AlCl_3 + 3H_2$$
$$Fe_2O_3 + 3H_2 = 2Fe + 3H_2O$$
$$O_2 + 2H_2 = 2H_2O$$
$$Ca + 2H_2O = Ca\,(OH)_2 + H_2$$

577. What role (that of oxidizing or reducing agent) does the oxygen play in the reactions expressed by the following equations:

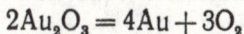

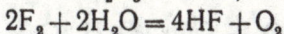

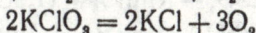

$$2Au_2O_3 = 4Au + 3O_2$$
$$2F_2 + 2H_2O = 4HF + O_2$$
$$2KClO_3 = 2KCl + 3O_2$$

Describe the electron transfer in these reactions.

578. By analyzing the atomic structure, decide if sodium atoms or ions, oxygen ions, iodine atoms, and aluminium ions can play the role of the oxidizing agents. Explain your answers.

579. By examining the atomic structure, decide if the ions given below can play the role of the reducing agents:

$$Sn^{2+}, \qquad Cl^-, \qquad Ag^+, \qquad Al^{3+}, \qquad Zn^{2+}, \qquad Ti^{3+}$$

Justify your answer.

580. Which of the following ions can and which cannot play the role of the oxidizing agent:

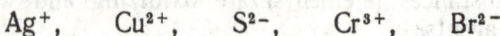

$$Ag^+, \qquad Cu^{2+}, \qquad S^{2-}, \qquad Cr^{3+}, \qquad Br^{2-}$$

Justify your answer.

581. Balance the following schemes:

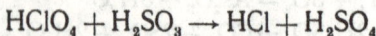

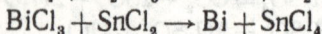

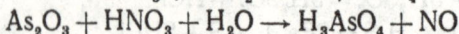

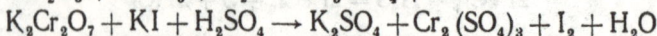

$$HClO_4 + H_2SO_3 \longrightarrow HCl + H_2SO_4$$
$$BiCl_3 + SnCl_2 \longrightarrow Bi + SnCl_4$$
$$As_2O_3 + HNO_3 + H_2O \longrightarrow H_3AsO_4 + NO$$
$$K_2Cr_2O_7 + KI + H_2SO_4 \longrightarrow K_2SO_4 + Cr_2(SO_4)_3 + I_2 + H_2O$$

Specify for each reaction: (a) what is oxidized and what is reduced; (b) what substance (and which atom or ion particularly) is the oxidizing and which is the reducing agent; (c) how the valence of the oxidizing and reducing agent changes in the reaction.

582. Make out balanced equations for the following reactions:

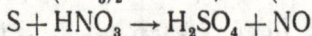

$$Zn + Pb(NO_3)_2 \longrightarrow Pb + Zn(NO_3)_2$$
$$S + HNO_3 \longrightarrow H_2SO_4 + NO$$
$$Cr_2O_3 + KNO_3 + KOH \longrightarrow K_2CrO_4 + KNO_2 + H_2O$$

*583. Make out balanced equations for the following reactions:

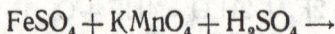

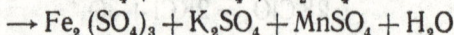

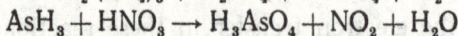

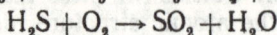

$$FeSO_4 + KMnO_4 + H_2SO_4 \longrightarrow$$
$$\longrightarrow Fe_2(SO_4)_3 + K_2SO_4 + MnSO_4 + H_2O$$
$$AsH_3 + HNO_3 \longrightarrow H_3AsO_4 + NO_2 + H_2O$$
$$H_2S + O_2 \longrightarrow SO_2 + H_2O$$

---

* While solving the problems marked with an asterisk, answer also the questions put in problem 581.

*584. Make out balanced equations for the following reactions:

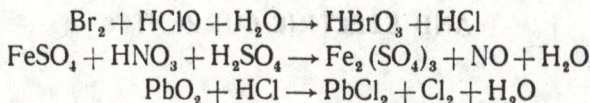

$$Br_2 + HClO + H_2O \rightarrow HBrO_3 + HCl$$
$$FeSO_4 + HNO_3 + H_2SO_4 \rightarrow Fe_2(SO_4)_3 + NO + H_2O$$
$$PbO_2 + HCl \rightarrow PbCl_2 + Cl_2 + H_2O$$

*585. Make out balanced equations for the following reactions:

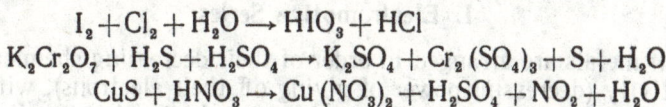

$$I_2 + Cl_2 + H_2O \rightarrow HIO_3 + HCl$$
$$K_2Cr_2O_7 + H_2S + H_2SO_4 \rightarrow K_2SO_4 + Cr_2(SO_4)_3 + S + H_2O$$
$$CuS + HNO_3 \rightarrow Cu(NO_3)_2 + H_2SO_4 + NO_2 + H_2O$$

*586. Make out balanced equations for the following reactions:

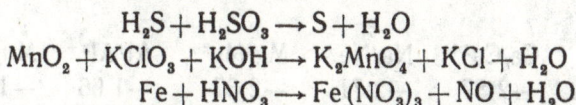

$$H_2S + H_2SO_3 \rightarrow S + H_2O$$
$$MnO_2 + KClO_3 + KOH \rightarrow K_2MnO_4 + KCl + H_2O$$
$$Fe + HNO_3 \rightarrow Fe(NO_3)_3 + NO + H_2O$$

*587. Make out balanced equations for the following reactions:

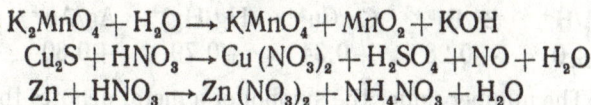

$$K_2MnO_4 + H_2O \rightarrow KMnO_4 + MnO_2 + KOH$$
$$Cu_2S + HNO_3 \rightarrow Cu(NO_3)_2 + H_2SO_4 + NO + H_2O$$
$$Zn + HNO_3 \rightarrow Zn(NO_3)_2 + NH_4NO_3 + H_2O$$

*588. Make out balanced equations for the following reactions:

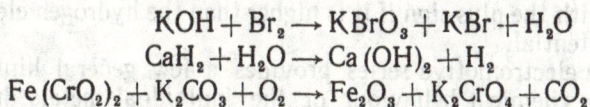

$$KOH + Br_2 \rightarrow KBrO_3 + KBr + H_2O$$
$$CaH_2 + H_2O \rightarrow Ca(OH)_2 + H_2$$
$$Fe(CrO_2)_2 + K_2CO_3 + O_2 \rightarrow Fe_2O_3 + K_2CrO_4 + CO_2$$

*589. Make out balanced equations for the following reactions:

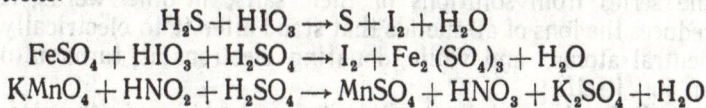

$$H_2S + HIO_3 \rightarrow S + I_2 + H_2O$$
$$FeSO_4 + HIO_3 + H_2SO_4 \rightarrow I_2 + Fe_2(SO_4)_3 + H_2O$$
$$KMnO_4 + HNO_2 + H_2SO_4 \rightarrow MnSO_4 + HNO_3 + K_2SO_4 + H_2O$$

CHAPTER XVI

CHEMICAL PROCESSES AND ELECTRIC CURRENT

1. Electromotive Series

If metals are arranged in order of their decreasing chemical activity (decreasing power of giving off their electrons), with hydrogen included in this series, we obtain the so-called *displacement series* or the *electromotive series* of metals. The electromotive series for the most important metals is as follows:

K/K^+	Ca/Ca^{2+}	Na/Na^+	Mg/Mg^{2+}	Al/Al^{3+}	Mn/Mn^{2+}
—2.92	—2.87	—2.71	—2.37	—1.66	—1.18

Zn/Zn^{2+}	Fe/Fe^{2+}	Ni/Ni^{2+}	Sn/Sn^{2+}	Pb/Pb^{2+}
—0.76	—0.44	—0.25	—0.14	—0.13

H_2/H^+	Bi/Bi^{3+}	Cu/Cu^{2+}	Hg/Hg_2^{2+}	Ag/Ag^+	Au/Au^{3+}
±0	+0.28	+0.34	+0.79	+0.80	+1.50

The number under the symbol of a metal denotes the normal potential of the metal in volts related to the normal hydrogen electrode whose potential is assumed zero.

The potential is marked with the minus sign if it is lower and with the plus sign if it is higher than the hydrogen electrode potential.

The electromotive series provides a few general hints as to the chemical behaviour of the individual metals during reactions:

1. Each metal displaces all the metals standing after it in the series from solutions of their salts, in other words, it reduces the ions of all metals that stand after it to electrically neutral atoms, and while donating electrons it turns into an ion itself.

2. Only the metals standing before hydrogen in the electromotive series can displace it from acid solutions (for example, Zn, Fe, Sn, but not Cu or Hg).

3. The farther to the left a metal is in the electromotive series, i.e. the lower its potential, the more active it is, the greater its reducing capacity with respect to ions of other metals, the easier it turns into an ion and the more difficult its ion is to reduce.

2. Galvanic Cells

Any pair of metals immersed into solutions of their salts which communicate through a porous partition, or through a syphon filled with an electrolyte, forms a *galvanic cell*. Metal plates immersed into solutions are called the *electrodes* of the cell.

If the outer ends of the electrodes (electrode poles) are connected by a metal wire, an electric current immediately arises, the electrons being transferred from the metal in which the algebraic value of the potential is less toward the metal in which it is greater (for example, from zinc to nickel, from nickel to copper, etc.). The movement of electrons upsets the equilibrium between the metal and its ions in the solution, and new ions pass into the solution. The metal is gradually dissolving. Simultaneously, the electrons transferred to the other metal discharge ions in the solution at its surface. As a result, the metal is deposited from the solution. The former metal is the negative electrode of the cell and the latter is the positive electrode. The negative electrode in a galvanic cell is called the anode and the positive the cathode.

In other words, in a closed galvanic cell, a metal reacts with a solution of a salt of another metal, which are separated from each other. Atoms of the first metal donate electrons to turn into ions, whereas the ions of the other metal accept the electrons to turn into atoms. The first metal displaces the second from the solution of its salt. For example, during operation of a galvanic cell comprising zinc and lead plates immersed into normal solutions of zinc nitrate and lead nitrate respectively, the following processes take place at the electrodes:

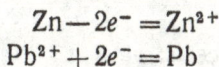

$$Zn - 2e^- = Zn^{2+}$$
$$Pb^{2+} + 2e^- = Pb$$

Summing up the two processes we get the ionic equation of

the reaction taking place in the cell:

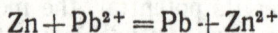

$$Zn + Pb^{2+} = Pb + Zn^{2+}$$

The molecular equation of the same reaction is

$$Zn + Pb(NO_3)_2 = Pb + Zn(NO_3)_2$$

The electromotive force of a galvanic cell is equal to the difference of the potentials of its two electrodes. In determining the electromotive force, the lower potential is always subtracted from the higher one. For example, the electromotive force (emf) of the galvanic cell discussed above is $-0.13-(-0.76)=0.63$ V. This will be the magnitude of its electromotive force provided the metals are immersed into the solutions in which the concentrations of their ions are 1 g-ion/litre. With other concentrations the electrode potentials will be different, and can be calculated from the formula:

$$E = E_0 + \frac{0.058}{n} \log C$$

where E is the sought potential of metal in volts, E_0 is its normal potential, n is the valence of metal ions and C is the concentration of ions in the solution expressed in gram-ions per litre.

Example. Find the electromotive force of a cell formed by a zinc electrode immersed in a $0.1M$ solution of zinc nitrate $Zn(NO_3)_2$ and a lead electrode immersed into a $2M$ solution of lead nitrate $Pb(NO_3)_2$.

Solution. First calculate the potential of the zinc electrode:

$$E_{Zn} = -0.76 + \frac{0.058}{2} \log 0.1 = -0.76 + 0.029(-1) = -0.79\,V$$

Calculate now the potential of the lead electrode:

$$E_{Pb} = -0.13 + \frac{0.058}{2} \log 2 = -0.13 + 0.029 \times 0.3010 = -0.12\,V$$

The electromotive force is:

$$emf = -0.12 - (-0.79) = 0.67\ V$$

PROBLEMS

590. Nickel plates are immersed into solutions of the following salts:

$$MgSO_4, \quad NaCl, \quad CuSO_4, \quad AuCl_3, \quad ZnCl_2, \quad Pb(NO_3)_2$$

With what salt will nickel react? Write the molecular and ionic equations for the reactions.

591. The following solutions are filled into six test tubes:

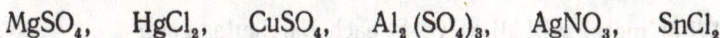

$$MgSO_4, \quad HgCl_2, \quad CuSO_4, \quad Al_2(SO_4)_3, \quad AgNO_3, \quad SnCl_2$$

A piece of zinc is placed into each test tube. With what solution will the zinc react? Write the molecular and ionic equations for the reactions.

592. Which pair of the substances given below will react?

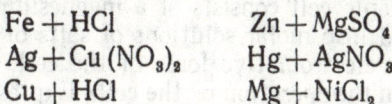

$$Fe + HCl \qquad\qquad Zn + MgSO_4$$
$$Ag + Cu(NO_3)_2 \qquad Hg + AgNO_3$$
$$Cu + HCl \qquad\qquad Mg + NiCl_2$$

Express the reactions by the molecular and ionic equations.

593. In which pairs of the substances given below will the reaction of displacement take place?

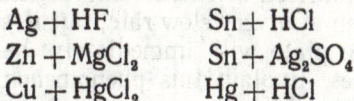

$$Ag + HF \qquad\qquad Sn + HCl$$
$$Zn + MgCl_2 \qquad\quad Sn + Ag_2SO_4$$
$$Cu + HgCl_2 \qquad\quad Hg + HCl$$

Write the ionic equations for the reactions.

594. An iron plate was immersed into a solution of blue vitriol. As soon as copper deposited on the plate it was removed from the solution, rinsed, dried and weighed. Its mass increased 2 grams. How many grams of copper deposited on the plate?

595. A copper plate weighing 50 g was immersed into a solution of mercuric chloride $HgCl_2$. On termination of the reaction the plate was extracted from the solution, washed, dried and weighed. The new weight was 52.74 g. How many grams of $HgCl_2$ did the solution contain?

596. Nickel and cadmium plates were placed in a solution of sulphuric acid. When the outer ends of the plates were interconnected with a wire, hydrogen began to evolve at the surface of the nickel plate. Can we decide which of the metals

7*

stands farther to the left in the electromotive series? Express
the reaction by an equation.

597. What will be the direction of electron current in the
wire connecting the following pairs in galvanic cells:

(1) $Mg/Mg(NO_3)_2 - Pb/Pb(NO_3)_2$
(2) $Pb/Pb(NO_3)_2 - Cu/Cu(NO_3)_2$
(3) $Cu/Cu(NO_3)_2 - Ag/AgNO_3$

What metal will dissolve in each particular case?

598. Make two cells, in one of which copper is the cathode
and in the other it is the anode. Write the equations for the
reactions which take place during operation of the cells and
indicate approximately their electromotive forces.

599. A galvanic cell consists of a magnesium and an iron
plate immersed into molar solutions of salts of these metals.
Determine the electromotive force of the cell. Which metal
will be spent in the operation of the cell? Express the reaction
by the molecular and ionic equations.

600. How can iron displace copper from its sulphate solution without immersing iron into this solution?

601. When immersed into hydrochloric acid an iron plate
liberates hidrogen at a very slow rate, but if we touch it with
a zinc wire, the plate will immediately be covered with
hydrogen bubbles. Explain this phenomenon. Which metal
transfers into the solution?

602. A galvanic cell comprises a silver electrode immersed
into a 1M solution of silver nitrate and a normal hydrogen
electrode. What chemical processes will take place at the electrodes of the cell during its operation? Find the electromotive force of the cell.

603. When placed into dilute sulphuric acid, an iron or
a zinc plate is dissolved with evolution of hydrogen. What
happens if we immerse both plates in the acid simultaneously
and connect their outer ends with a metal wire? Will both
of them dissolve? At the surface of which plate will hydrogen
be evolved? Justify your answers.

604. One zinc electrode is immersed into a dilute solution
of zinc sulphate, and the other zinc electrode into a concentrated solution of the same salt. The solutions communicate through a tube filled with a solution of sodium sulphate. An elect-

ric current is generated if the electrodes are connected with a metal wire. Explain this phenomenon and write electronic equations for the reactions which take place at the electrodes.

605. What chemical reactions take place at the electrodes during operation of a galvanic cell consisting of copper and silver plates immersed into molar solutions of copper nitrate and silver nitrate? Indicate the emf of the cell.

606. Calculate the emf of a cell formed by a nickel electrode immersed into a $0.1M$ solution of nickel sulphate and a copper electrode immersed into a $0.2M$ solution of copper sulphate, if the dissociation of the salts is assumed to be complete. Express the reaction that takes place in the working cell by the molecular and ionic equations.

607. What process will take place at the electrodes of a working galvanic cell consisting of aluminium and silver plates immersed into a $0.01M$ solution of an aluminium salt and a $2M$ solution of silver nitrate respectively? What is the emf of the cell if the dissociation of the salts is assumed to be complete?

608. What is the potential of the hydrogen electrode immersed into pure water, if the normal potential of the hydrogen electrode is assumed to be zero?

609. The electromotive force of a cell consisting of copper and lead electrodes immersed into molar solutions of the respective salts is 0.47 V. Will the emf change if $0.001M$ solutions are taken? Justify your answer by the appropriate calculations.

610. In what direction will the electrons flow in a wire connecting the poles of a cell consisting of a tin electrode immersed into a molar solution of tin chloride $SnCl_2$ and a hydrogen electrode immersed into a solution containing 10^{-4} gram-ion of hydrogen per litre? Which electrode of this cell is negative?

3. Electrolysis

Electrolysis is the process of decomposition of a substance with an electric current.

The essence of electrolysis is that when an electric current is passed through a solution of an electrolyte (or through the molten electrolyte), positively charged ions move toward the cathode, and the negative ions toward the anode, where they are discharged. As a result, at the electrodes the component parts of the dissolved electrolyte are liberated or hydrogen and oxygen are evolved from water.

Some ions lose their charges more easily than others. Different voltages are therefore required to convert various ions into neutral atoms or groups of atoms. The degree of easiness at which metal ions are discharged (that is gain electrons) is determined by the position of a particular metal in the electromotive series. The farther to the left the metal stands in the series and the greater its negative potential (or the lesser its positive potential), the more difficult it gains the electrons. When comparing metals in the electromotive series, one can see that trivalent ions of gold, then silver ions, etc., will be discharged more easily (with the least electric current voltages); potassium ions will be discharged with the greatest difficulty (at the highest voltage).

If ions of two or more metals are present in a solution simultaneously, first will be discharged the ions of the metal whose negative potential is less (or the positive greater). For example, metallic copper is first liberated from a solution containing ions of zinc and copper. But the potential of a metal is known to vary with the concentration of its ions in the solution. In exactly the same way the easiness at which ions of each metal are discharged changes with their concentration: ions are discharged more easily at higher concentrations and with lower concentrations the discharge is more difficult. It may happen therefore that, during electrolysis of a solution containing ions of several metals, the more active metal will be deposited sooner than the other, less active metal (if the concentration of the first metal ions is significant and that of the second is very small).

In addition to salt ions, aqueous solutions of salts always contain ions of water (H^+ and OH^- ions).

Although theoretically hydrogen ions should be the first to discharge (compared with the metals standing before hydrogen in the electromotive series), practically, the metals are deposited at the cathode during electrolysis of all salts, except those of the most active metals, owing to the insignificant concentration of hydrogen ions. It is only during electrolysis of salts of sodium, calcium and other metals (up to aluminium and including), that hydrogen ions are discharged and hydrogen gas is evolved.

Ions of acid radicals or hydroxyl ions of water are discharged at the anode.

If ions of acid radicals do not contain oxygen (for example, ions Cl^-, S^{2-}, CN^- and others), these very ions (and not the hydroxyl ions) are discharged, as the latter give up their charge with much greater difficulty, the result being that chlorine, sulphur, etc., are liberated at the anode. If, however, a salt of an oxygen-containing acid, or the acid itself, is subject to electrolysis, hydroxyl ions, and not the ions of the acid radicals, are discharged.

When hydroxyl ions are discharged, they turn into electrically neutral OH groups, which immediately turn into water and oxygen molecules:

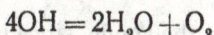

$$4OH = 2H_2O + O_2$$

Oxygen is liberated at the anode as a result.

Below follow a few examples of electrolysis.

1. *Electrolysis of nickelous chloride* $NiCl_2$ *solution*. The solution contains Ni^{2+} and Cl^- ions. As electric current is passed through the solution, the nickel ions move toward the cathode, and the chloride ions toward the anode. Each nickel ion accepts two electrons from the cathode and turns into electrically neutral atoms which are deposited on the cathode surface.

Chloride ions, as they reach the anode, give up electrons and turn into chlorine atoms, which combine in pairs to form molecules of chlorine. Free chlorine gas is evolved at the anode as a result.

Schematically this process can be represented as follows:

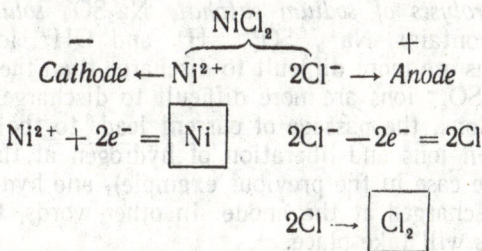

$$Cathode \leftarrow Ni^{2+} \quad \overbrace{NiCl_2} \quad 2Cl^- \rightarrow Anode$$

$$Ni^{2+} + 2e^- = \boxed{Ni} \qquad 2Cl^- - 2e^- = 2Cl$$

$$2Cl \rightarrow \boxed{Cl_2}$$

Thus, the reduction takes place at the cathode and oxidation at the anode.

2. *Electrolysis of potassium iodide* KI *solution.* Potassium iodide is present in solution in the form of K^+ and I^- ions. As electric current is passed through, potassium ions move toward the cathode, and iodide ions toward the anode. But potassium stands much farther to the left than hydrogen in the electromotive series, therefore these are hydrogen ions of the water and not the potassium ions that are discharged at the cathode. Hydrogen atoms combine in pairs to form hydrogen molecules which are liberated at the cathode.

As the hydrogen ions are discharged, new molecules of water dissociate; as a result, hydroxyl ions (released from the water molecules), and also potassium ions are accumulated at the cathode and a solution of potassium hydroxide is produced as a result.

Iodine is liberated at the anode, since iodide ions are discharged more easily than the hydroxyl ions of water:

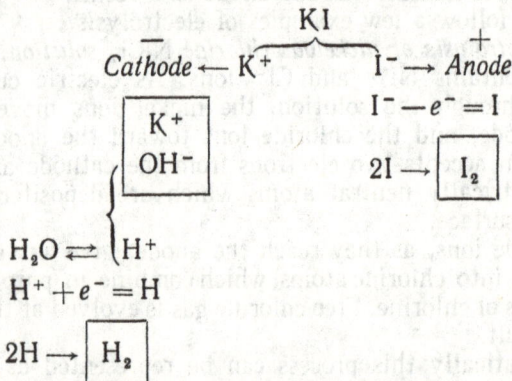

$$\overset{\displaystyle \overbrace{\qquad\qquad}^{\text{KI}}}{Cathode \leftarrow \overset{-}{K^+} \qquad \overset{}{I^-} \longrightarrow \overset{+}{Anode}}$$

$$I^- - e^- = I$$

$$\boxed{K^+}$$
$$\boxed{OH^-}$$

$$2I \longrightarrow \boxed{I_2}$$

$$H_2O \rightleftarrows \begin{cases} H^+ \\ \\ H^+ + e^- = H \end{cases}$$

$$2H \longrightarrow \boxed{H_2}$$

3. *Electrolysis of sodium sulphate* Na_2SO_4 *solution.* The solution contains Na^+, SO_4^{2-}, H^+ and OH^- ions. Since sodium ions are more difficult to discharge than the hydrogen ions, and SO_4^{2-} ions are more difficult to discharge than the hydroxyl ions, the passage of current leads to the discharge of hydrogen ions and liberation of hydrogen at the cathode (as was the case in the previous example), and hydroxyl ions will be discharged at the anode. In other words, the water electrolysis will take place.

At the same time, owing to the discharge of the hydrogen and hydroxyl ions of water and the continuous movement of the sodium ions toward the cathode, and SO_4^{2-} ions toward the anode, a solution of alkali (NaOH) forms at the cathode and a solution of acid (H_2SO_4) at the anode:

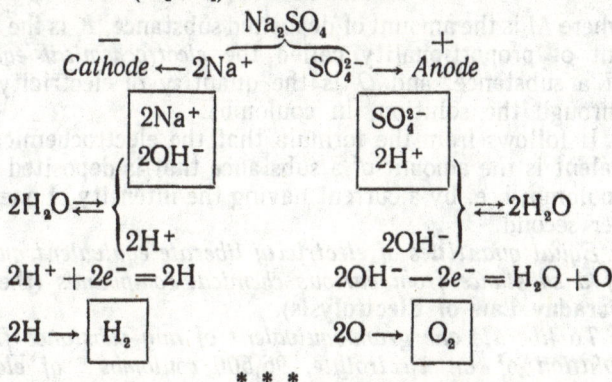

$$\overset{-}{Cathode} \leftarrow 2Na^+ \quad \overbrace{Na_2SO_4} \quad SO_4^{2-} \rightarrow \overset{+}{Anode}$$

$$2H_2O \rightleftharpoons \left\{ \begin{matrix} 2Na^+ \\ 2OH^- \\ 2H^+ \end{matrix} \right. \qquad \left. \begin{matrix} SO_4^{2-} \\ 2H^+ \\ 2OH^- \end{matrix} \right\} \rightleftharpoons 2H_2O$$

$$2H^+ + 2e^- = 2H \qquad \qquad 2OH^- - 2e^- = H_2O + O$$

$$2H \rightarrow \boxed{H_2} \qquad \qquad 2O \rightarrow \boxed{O_2}$$

* * *

There is a special case of electrolysis when the anode is made of the metal whose salt is present in the solution. No ions are discharged at the anode in this case, but the anode itself dissolves gradually yielding ions into the solution and giving up its electrons to the source of current.

Electrolysis of a solution of copper sulphate $CuSO_4$ with a copper anode may serve a good example of this specific case:

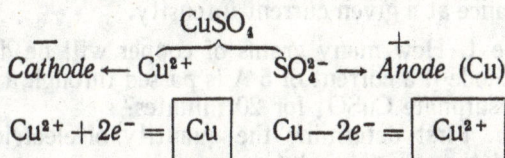

$$\overset{-}{Cathode} \leftarrow Cu^{2+} \quad \overbrace{CuSO_4} \quad SO_4^{2-} \rightarrow \overset{+}{Anode} \text{ (Cu)}$$

$$Cu^{2+} + 2e^- = \boxed{Cu} \qquad Cu - 2e^- = \boxed{Cu^{2+}}$$

During the process, copper is deposited on the cathode and the anode is gradually dissolved. The quantity of copper sulphate in the solution remains unaltered.

4. Laws of Electrolysis

The electrolytic processes obey the laws formulated in the thirties of last century by the English physicist Michael Faraday.

The weight of a substance deposited by electrolysis is proportional to the quantity of electricity passing through the solution and is quite independent of any other factors (the **first Faraday Law of Electrolysis**):

$$M = KQ$$

where M is the amount of deposited substance, K is the coefficient of proportionality called the *electrochemical equivalent* of a substance, and Q is the quantity of electricity passed through the solution, in coulombs.

It follows from the formula that the electrochemical equivalent is the amount of a substance that is deposited by one coulomb, i. e. by a current having the intensity of one ampere per second.

Equal quantities of electricity liberate equivalent quantities of a substance from various chemical compounds (the **second Faraday Law of Electrolysis**).

To liberate one gram-equivalent of any substance from the solution of an electrolyte, 96,500 coulombs * of electricity must be passed through the solution.

Once the Faraday laws are known, many important calculations connected with the electrolysis can be made. For example, it is possible (1) to calculate the quantity of a substance liberated or decomposed by a certain quantity of electricity; (2) to determine the current intensity from the quantity of a substance liberated and from the time spent for its liberation; (3) to establish the time required to liberate a certain quantity of a substance at a given current intensity.

Example 1. How many grams of copper will be deposited on the cathode if a current of 5 A is passed through a solution of copper sulphate $CuSO_4$ for 20 minutes?

Solution. First determine the quantity of electricity that has passed through the solution:

$$Q = it$$

where i is the current intensity, in amperes, and t is the time, in seconds.

In accordance with the conditions of the problem, $i=5A$, $t=20$ minutes or 1,200 seconds, whence:

$$Q = 5 \times 1,200 = 6,000 \text{ coulombs}$$

_____

* To be more exact, 96,491 coulombs

The equivalent weight of copper (atomic weight, 63.54) is $63.54:2=31.77$. It follows therefore that 96,500 coulombs will liberate 31.77 g of copper. The sought quantity of copper will then be

$$M = \frac{31.77 \times 6,000}{96,500} = 1.975 \text{ g}$$

Example 2. How long shall electric current of 10A be passed through a solution of an acid in order to liberate 5.6 litres of hydrogen at STP?

Solution. Let us find the quantity of electricity that should be passed through the solution of an acid to liberate 5.6 litres of hydrogen. Since the gram-equivalent of hydrogen at STP occupies the volume of 11.2 litres, the sought quantity of electricity is

$$Q = \frac{96,500 \times 5.6}{11.2} = 48,250 \text{ coulombs}$$

Now the time during which the current should be passed can be determined:

$t = \frac{Q}{i} = \frac{48,250}{10} = 4,825$ seconds, or 1 hour, 20 minutes and 25 seconds.

Example 3. Current passed through a solution of a silver salt for ten minutes liberated 1 g of silver. What was the current intensity?

Solution. One gram-equivalent of silver is 107.9 g. To liberate 1 g of silver, $96,500:107.9=894$ coulombs should pass through the solution. Hence the current intensity is

$$i = \frac{894}{10 \times 60} \approx 1.5 \text{ A}$$

Example 4. What is the equivalent weight of tin if 2.77 g of the metal is liberated from a solution of tin chloride $SnCl_2$ by passing electric current of 2.5A for 30 minutes?

Solution. The quantity of electricity passed through the solution during 30 minutes is

$$Q = 2.5 \times 30 \times 60 = 4,500 \text{ coulombs}$$

Since 96,500 coulombs are required to liberate one gram-equivalent of any substance, the equivalent weight of tin is:

$$\text{Equivalent weight of tin} = \frac{2.77 \times 96{,}500}{4{,}500} = 59.4$$

Example 5. What quantity in grams of zinc should be dissolved per minute in the galvanic cell $Zn/ZnSO_4$—$Cu/CuSO_4$ in order to ensure current intensity of 3 A?

Solution. Zinc is a bivalent metal. Hence its one gram-equivalent (32.7 g) carries 96,500 coulombs. To ensure the current intensity of 3 A, $3 \times 60 = 180$ coulombs should be passed through the galvanic cell per minute. Hence, the quantity of zinc that should be dissolved during one minute is

$$\frac{32.7 \times 180}{96{,}500} = 0.061 \text{ g}$$

PROBLEMS

611. Which metal will be the first to deposit from the solution containing sulphates of nickel, silver and copper, if the current density is sufficient to liberate either of these metals?

612. A solution contains Fe^{2+}, Hg_2^{2+}, Bi^{3+} and Pb^{2+} ions in equal concentrations. In what order will these ions be discharged during electrolysis?

613. Can any metal be liberated by electrolyzing an aqueous solution of its salt?

614. Make out schemes for electrolysis of aqueous solutions of barium chloride and lead nitrate.

615. What oxidation and reduction processes take place during electrolysis of aqueous solutions of ferric chloride and calcium nitrate?

616. What processes take place at the cathode and the anode during electrolysis of a zinc chloride solution, if the anode is (a) carbon and (b) zinc?

617. Electric current is passed through solutions of potassium carbonate and potassium sulphate. Will the same substances be liberated from the solutions?

618. According to the modern views, the charge on the electron is 1.6×10^{-19} coulomb. Determine the Faraday unit (faraday).

619. Through solutions of sodium chloride and trisodium phosphate electric current was passed for a certain period of time. Have the quantities of the salts changed in the solutions?

620. What chemical processes take place at the anode and the cathode during electrolysis of a solution of potassium nitrate if both electrodes are copper?

621. During electrolysis of a sulphate solution, 176 ml of oxygen at STP were evolved at the anode, and 1 gram of metal was liberated at the cathode. What is the atomic weight of the metal?

622. Two zinc plates are immersed into a vessel containing a solution of sulphuric acid so that they do not touch each other. Both plates dissolve in the acid with liberation of hydrogen. Now the plates are connected to the poles of a source of electricity. What chemical processes will take place in the vessel?

623. Make out a scheme of electrolysis of an aqueous solution of magnesium sulphate. What is oxidized and what is reduced in this process?

624. A current of 3 A was passed through a solution of ferrous chloride $FeCl_2$ for ten minutes and a current of 5 A through a solution of ferric chloride $FeCl_3$ for six minutes. In which case is the weight of the deposited iron greater? Justify your answer.

625. Electric current was passed for a period of time through solutions of stannous chloride $SnCl_2$ and stannic chloride $SnCl_4$ connected in series. Were the quantities of tin and chlorine liberated from the two solutions equal?

626. Current was passed through a solution of nickel sulphate $NiSO_4$ until all nickel was liberated from the solution. What processes took place at the anode and the cathode? What is the liquid that remained after the process?

627. Equal quantities of electricity were passed through solutions of silver nitrate $AgNO_3$ and bismuth nitrate $Bi(NO_3)_3$, 0.9 g of silver was deposited on the cathode from the first solution. How many grams of bismuth were liberated on the cathode from the second solution?

628. During electrolysis of a solution of chromium nitrate $Cr(NO_3)_3$, 0.26 g of chromium was liberated at the cathode during 10 minutes. What was the current intensity?

629. What is the intensity of current that will liberate 280 ml of fire damp, measured at STP, from a solution of sulphuric acid during 20 minutes?

630. How long will it take to decompose one gram-molecule of water with a current of 5 A?

631. How long shall the current of 3 A be passed through a solution of a silver salt to coat a surface of 80 sq cm with a layer 0.005 mm thick, if the density of silver is 10.5 g/cu cm?

632. What quantity of potassium hydroxide will be liberated at the cathode if 9,650 coulombs are passed through a solution of a potassium salt?

633. How much electricity must traverse a solution of a silver salt in order to liberate 1 g of silver?

634. How much sulphuric acid is formed in a solution of copper sulphate if 1,930 coulombs are passed through the solution?

635. What processes will take place at the anode and the cathode during electrolysis of a solution of a nickel salt, if both electrodes are nickel? How will the mass of the anode change after a current of 2.5 A has been passed for an hour?

636. What oxidation and reduction processes will take place at the electrodes during electrolysis of a solution of cadmium sulphate $CdSO_4$? What substances and in what quantities will be liberated at the electrodes after the passage of a current of 4 A for 40 minutes?

637. A cell with a zinc anode produced a current of 0.8 A for two hours. What quantity of zinc was spent?

638. A solution of magnesium chloride was electrolyzed for an hour with a current of 2.5 A. What ions were discharged at the cathode and the anode? What substances and in what quantities were liberated at the anode and the cathode?

639. Make out a scheme of electrolysis of a solution of sodium chloride and calculate how many coulombs traverse the solution during formation of 1 kg of sodium hydroxide.

640. A solution of barium iodide BaI_2 was electrolyzed for 15 minutes at a current intensity of 6 A. Make out a scheme of electrolysis and calculate what quantities of substances were liberated from the solution.

641. A current of 5 A liberates 1.517 g of platinum from a solution of its salt during 10 minutes. What is the equivalent weight of platinum?

642. What is the gram-equivalent of bismuth if 13,850 coulombs are required to liberate 10 grams of bismuth from a solution of its salt?

643. Determine the gram-equivalent of cadmium, knowing that 1 g of cadmium can be liberated by passing 1,717 coulombs through a solution of its salt.

644. As a current of 1.5 A traverses a solution of a salt of a trivalent metal during 30 minutes, 1.07 g of the metal are liberated at the cathode. What is the atomic weight of the metal?

645. Calculate the electrochemical equivalent of copper (atomic weight of copper is 63.54).

646. During electrolysis of a sulphate, 3.49 g of the metal were liberated at the cathode and 0.7 litre of oxygen was evolved at the anode. What is the exact atomic weight of the metal if its specific heat is 0.11?

647. Potassium chlorate is prepared by electrolyzing a hot solution of potassium chloride. Write the equations for all reactions that take place during this process and calculate how much potassium chlorate will be produced if 193,000 coulombs are passed through the solution. Assume the yield to be 60 per cent.

648. Calculate the electrochemical equivalent of silver knowing that its atomic weight is 107.88.

CHAPTER XVII

ALLOYS

Metal alloys are mixtures of metal crystals or (if the metals react with each other) mixtures of free metals with their chemical compounds.

Solid alloys are sometimes quite homogeneous; they are either definite chemical compounds or homogeneous mixtures of indefinite composition known as solid solutions.

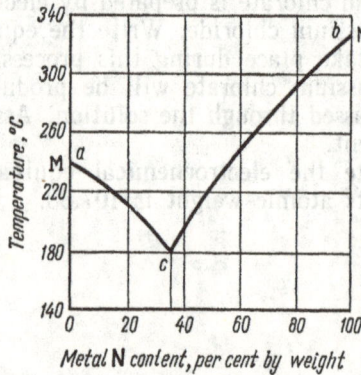

Fig. 3

The nature of alloys would be usually determined by the thermal analysis, which consists in plotting a phase diagram showing the relationship between the melting point of an alloy and its composition.

If two metals M and N do not form chemical compounds on alloying, the phase diagram has the appearance as in Fig. 3. Point *a* on the curve *acb* is the melting point of pure metal M. As more and more metal N is added to it, the melting point drops gradually until it reaches a certain point *c*, after which, if the metal N content is still increased, the melting point rises again until it reaches point *b* which is the melting point of pure metal N.

The curve *acb* shows that of all alloys that can be formed by metals M and N, that alloy will have the lowermost melting point, which corresponds to the point *c* (in this particular case it contains 37 per cent of metal N and 63 per cent of metal M). The alloy having the lowest possible melting point is called the *eutectic mixture* or just the *eutectic*, and the temperature at which it melts is called the *eutectic temperature*. The point *c* on the curve, corresponding to the eutectic temperature, is called the *eutectic point*.

On cooling a liquid alloy having the composition other than eutectic, the metal whose content is greater than in the

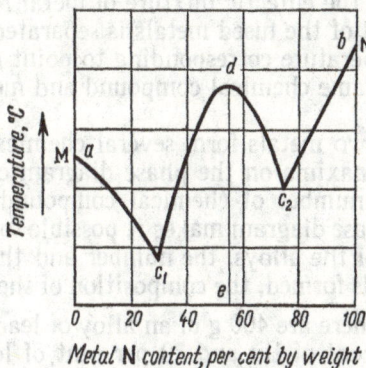

Metal N content, per cent by weight

Fig. 4

eutectic will crystallize out. For example, during cooling of an alloy containing 70 per cent of metal N, the latter will crystallize out first. As the metal N crystallizes, the temperature drops, and the composition of the remaining liquid phase of the alloy approaches gradually the composition of the eutectic. As soon as the composition of the liquid part of the alloy approaches that of the eutectic, and the temperature reaches the eutectic point, all of the liquid alloy will solidify to form a mixture of intimately intermixed finest crystals of both metals. Therefore, all alloys having the composition other than eutectic, when solid, are a mass of the eutectic into which larger crystals of the metal, which was separated first, are incorporated.

If the fused metals form only one chemical compound, the phase diagram has the appearance shown in Fig. 4. There are two eutectic points in the diagram, c_1 and c_2. The maximum on the curve (point d) corresponds to the melting point of the chemical compound formed by metals M and N, and the point e on the abscissa axis indicates its composition.

A perpendicular drawn from the point d to intersect the abscissa axis divides the area of the diagram into two parts that can be considered as two independent diagrams. The left-hand part of the curve is the phase diagram of metal M and the chemical compound formed by both metals, while the right-hand part of the curve is the phase diagram of the same chemical compound and metal N. At a temperature corresponding to point c_1 the eutectic mixture of metal M and the chemical compound of the fused metals is separated out from the alloy; at a temperature corresponding to point c_2 the eutectic mixture of the same chemical compound and metal N is separated.

If on fusing two metals form several chemical compounds, the number of maxima on the phase diagram curve will correspond to the number of chemical compounds produced.

Thus, the phase diagram makes it possible to determine the general nature of the alloys, the number and the composition of the compounds formed, the composition of the eutectic, etc.

Example 1. There are 400 g of an alloy of lead and tin, containing 30 per cent of tin and 70 per cent of lead. Which of the metals and in what quantity is incorporated as crystals into the mass of the eutectic if the latter contains 64 per cent of tin and 36 per cent of lead?

Solution. First calculate the number of grams of each metal contained in the 400 g of the alloy:

$$400 \times 0.30 = 120 \text{ g Sn}$$
$$400 \times 0.70 = 280 \text{ g Pb}$$

Since the percentage of tin in the alloy is less than in the eutectic, it is evident that all tin is present in the eutectic. Hence the mass of the eutectic is

$$120 : x = 64 : 100$$
$$x = \frac{120 \times 100}{64} = 187.5 \text{ g}$$

The rest of the alloy are crystals of lead incorporated into the eutectic. The mass of the crystals is

$$400 - 187.5 = 212.5 \text{ g}$$

Example 2. An intermetallic compound Mg_2Sn is formed on fusing tin with magnesium. In what proportion should these metals be taken in order to obtain the alloy containing 20 per cent of free magnesium?

Solution. First determine the percentage of magnesium and tin in the Mg_2Sn. We get 28.7 per cent of magnesium and 71.3 per cent of tin.

In accordance with the conditions of the problem, 100 g of the alloy should contain 20 g of magnesium and 80 g of Mg_2Sn. Now determine the number of grams of each metal in the 80 g of Mg_2Sn:

$$80 \times 0.287 = 23 \text{ g Mg}$$
$$80 \times 0.713 = 57 \text{ g Sn}$$

It follows therefore that to prepare 100 g of the required alloy, $23 + 20 = 43$ g of magnesium are needed for 57 g of tin, that is the weight ratio of tin to magnesium should be 57:43.

PROBLEMS

649. By using the phase diagram given in Fig. 5 for the system Cd—Bi, determine, which of these metals and at what

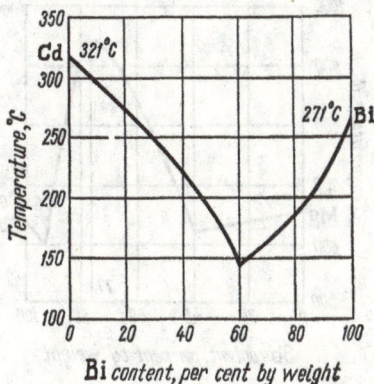

Fig. 5

temperature will separate first on cooling the following liquid alloys:

(a) 20 per cent Bi and 80 per cent Cd;

(b) 60 per cent Bi and 40 per cent Cd;

(c) 70 per cent Bi and 30 per cent Cd.

650. Which metal will separate on cooling a liquid alloy of copper and aluminium containing 25 per cent of copper, if the eutectic contains 67.5 per cent of aluminium and 32.5 per cent of copper? How many grams of this metal can be separated from 200 g of the alloy?

651. On cooling 500 g of a liquid alloy of copper and silver containing 77.6 per cent of silver, 100 g of pure silver were separated before the eutectic point was attained. Determine the percentage of the eutectic.

652. An alloy contains 73 per cent of tin and 27 per cent of lead. How many grams of the eutectic does 1 kg of the solid alloy contain, if the eutectic contains 64 per cent of tin and 36 per cent of lead?

653. Silver coins are usually minted out of an alloy containing equal weights of copper and silver. How many grams of crystalline copper are contained in 200 g of the alloy, if the eutectic contains 28 per cent of copper and 72 per cent of silver?

654. By using the phase diagram given in Fig. 6 for the system Mg—Sb, determine the formula of the intermetallic compound formed by these metals. What is the composition

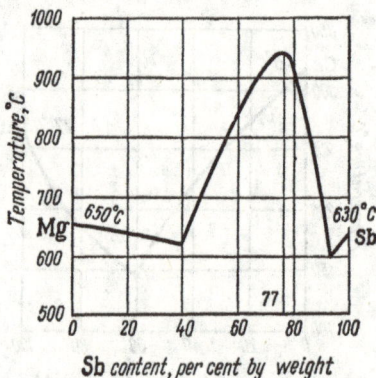

Fig. 6

of the solid phase that will separate the first on cooling the liquid alloy containing 60 per cent of antimony? What is the formula of the solidified alloy?

655. From the phase diagram given in Fig. 7 for the system Cu—Mg, determine the formulas of the intermetallic compounds formed by these metals.

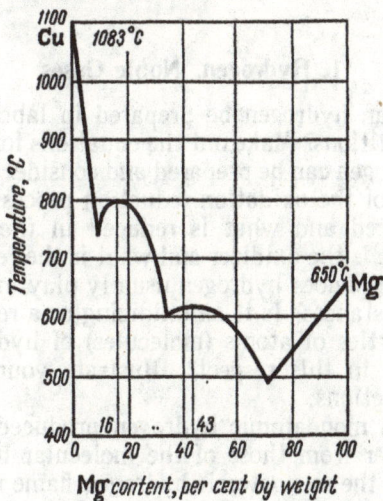

Fig. 7

656. An intermetallic compound containing 81 per cent of lead is formed by fusing magnesium and lead. Determine its formula and calculate the quantity in grams of this compound contained in 1 kg of an alloy formed by equal weights of magnesium and lead.

CHAPTER XVIII

PROPERTIES OF CHEMICAL ELEMENTS
AND THEIR COMPOUNDS

1. Hydrogen. Noble Gases

657. How can hydrogen be prepared in laboratory and in industrial conditions? Make out the equations for the reactions by which hydrogen can be prepared and consider them from the point of view of the oxidation-reduction process. Indicate (a) what is oxidized and what is reduced in the reaction, (b) what substance is the oxidizer and what is the reducing agent.

658. What role does hydrogen usually play in the reactions with other substances? Is it an oxidizing or a reducing agent? Are the properties of atoms (molecules) of hydrogen and its ions the same in this respect? Illustrate your answer with exemplary reactions.

659. How is monoatomic hydrogen produced? How do its properties differ from those of the molecular hydrogen? The temperature of the monoatomic hydrogen flame is much higher than that of the molecular hydrogen; why?

660. Derive the equation of the reaction employed in the manufacture of hydrogen by the steam-iron method. Despite the reversibility, the reaction continues practically to complete oxidation of iron; why? Indicate the main fields of practical uses of hydrogen.

661. Which of the metals given below can be used for the preparation of hydrogen from hydrochloric acid: copper, aluminium, iron, magnesium or mercury? When taken in equal quantities, which of them will displace the greatest amount of hydrogen?

662. How can metallic hydrides be produced? Derive equations for the reactions of (a) preparation of calcium hydride, (b) interaction of calcium hydride with water. Indicate, what substance is the oxidizing and what is the reducing agent. What is oxidized and what is reduced in both cases?

663. To fill aerostats with hydrogen in field conditions, the reaction between calcium hydride and water is often used. How many kilograms of the hydride are required to fill an aerostat having the capacity of 560 cu m (at STP)? How many kilograms of zinc and sulphuric acid are required for the same purpose?

664. Give the brief specifications for the noble gases, such as (a) their position in the Periodic System; (b) the atomic structure; (c) basic differences from the other elements; (d) the number of atoms in a molecule.

665. The density of helium is twice that of hydrogen. Calculate the weight (including that of the balloon) that can be lifted by an aerostat holding 1,000 cu m of (a) hydrogen; and (b) helium. Compare the lifting powers of hydrogen and helium.

666. Indicate the uses of neon. What is the mass of one litre of the gas at STP? What is its density with respect to air?

2. The Halogens

667. What are the basic halogen compounds? In the form of what ions do the halogens usually occur in natural compounds? Indicate the general principle used in the manufacture of the halogens from their compounds and illustrate your answer by the equation of the reaction employed in the preparation of bromine from sodium bromide.

668. Give the comparative characteristic of chemical properties of the halogens, indicating (a) their atomic structure; (b) the possible valence; (c) the formulas of their compounds with hydrogen and changes in the thermal stability with the increasing atomic number of the halogen.

How do the oxidizing and reducing properties of the halogens change with the increasing atomic number? What is the cause of these changes?

669. Indicate the methods for preparation and the physical properties of the halogen hydrides. Which of these compounds has the greatest practical importance? How do the reducing properties of the halogen hydrides change with the increasing atomic number of the halogen? Can halogen hydrides play the role of the oxidizing agents? Justify your answer.

670. How do the atoms (or molecules) of the halogens differ from the negatively charged halide ions with respect to the oxidizing-reducing properties? How do these properties change in both groups with the increasing atomic number of the halogen? What is the reason of these changes?

671. Consider the reactions given below from the oxidation-reduction point of view:

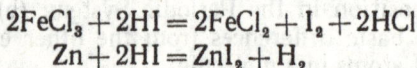

$$2FeCl_3 + 2HI = 2FeCl_2 + I_2 + 2HCl$$
$$Zn + 2HI = ZnI_2 + H_2$$

Does the hydriodic acid play the same role in both reactions? What substances are the reducing agents in the first reaction and in the second?

672. Name oxygen-containing acids of chlorine and their salts with calcium. Give their formulas. How do their oxidizing properties change with the growing valence of chlorine? What is the usual starting reaction in the preparation of the oxygen-containing compounds of chlorine?

673. How can chlorinated lime be produced from calcium carbonate and sodium chloride if a source of electricity is available? Indicate all chemical reactions that will take place during the preparation of chlorinated lime out of these substances. What will be the side products?

674. Write the chemical formula of potassium chlorate and indicate its practical uses. How can potassium chlorate be prepared if metallic potassium, hydrochloric acid, manganese dioxide and water are available? Make out the equations of all necessary reactions. How can the presence of potassium chloride admixture in a sample of potassium chlorate be detected?

675. Derive equations for the reactions which should be used to prepare potassium chlorate, if hydrochloric acid, potassium hydroxide, manganese dioxide and water are only available. Calculate the quantity of potassium chlorate that can be obtained from 168 g of potassium hydroxide.

676. By what reactions can bromine be obtained if sodium bromide, hydrochloric acid and manganese dioxide are available? Derive equations for these reactions and indicate what substance is the oxidizing and what is the reducing agent in each reaction.

677. By what process, other than electrolysis, can chlorine be obtained from calcium chloride? Derive equations for the reactions, and calculate how many litres of chlorine at STP can be obtained from 1 kilogram of calcium chloride.

678. A jet of chlorine gas was passed through a solution of potassium iodide for a lengthy period of time. The solution was then tested with starch for the presence of free iodine. The starch however did not colour blue. How can this fact be explained? Express this reaction by an equation.

679. How can iodic acid be prepared from manganese dioxide, hydrochloric acid and free iodine? Express the reactions by equations.

680. A laboratory method for preparing chlorine consists in the action of hydrochloric acid upon chlorinated lime. Explain the reaction from the oxidation-reduction viewpoint. How many litres of chlorine at STP can thus be obtained from 1 kilogram of chlorinated lime containing 42.9 per cent of calcium hypochlorite?

3. The Oxygen Family

681. Describe briefly the properties of oxygen, indicating (a) occurrence of oxygen and its content in the atmosphere; (b) position in the Periodic System, the atomic structure; (c) the oxidizing power and its sources.

Which elements cannot combine directly with oxygen and which do not combine with oxygen at all?

682. Into what groups are oxides divided? Indicate the basic signs of each group. Illustrate your answer with the appropriate reactions.

683. After a volume of oxygen has been ozonized, it was reduced 500 ml. How many millilitres of ozone were formed? What amount of energy was consumed in its formation? What volume of oxygen was spent for the formation of the ozone?

684. Indicate the laboratory and industrial methods for preparation of oxygen. In what state, liquid or gaseous, is oxygen stored in steel cylinders? Name the most important fields of practical uses of oxygen.

685. What pairs of the substances given below will react to form salts?

SO_2 and HCl; CaO and HNO_3; CrO_3 and $NaOH$;
MgO and KOH; P_2O_5 and CuO; N_2O_5 and H_3PO_4

Write the equations for these reactions.

686. A small quantity of manganese dioxide was added to 150 g of a solution of hydrogen peroxide. The oxygen evolved in the reaction was measured and recalculated with reference to standard conditions of temperature and pressure. The result was one litre. What had been the percentage content of hydrogen peroxide in the solution?

687. What compounds are known as peroxides? How can hydrogen peroxide be obtained from them? Derive the equation for the reaction of oxidation of ferrous hydroxide $Fe(OH)_2$ to ferric hydroxide $Fe(OH)_3$ by hydrogen peroxide, and indicate, which ion is the reducing and which is the oxidizing agent.

688. Express by the ionic equation the reaction of hydrolysis of sodium peroxide. How will the rising temperature affect the equilibrium in this reaction? Will the solution of sodium peroxide retain its bleaching properties if boiled?

689. Describe briefly the chemical properties of sulphur, indicating (a) its position in the Periodic System, the atomic structure, valence in various compounds; (b) oxidizing properties (exemplify them with an equation of the reaction); (c) properties of sulphur oxides; (d) properties of the sulphur compound with hydrogen and its oxidizing or reducing properties.

690. How can hydrogen sulphide be obtained from zinc, sulphur and sulphuric acid? Indicate two possible methods for its preparation (exemplify with equations of the reactions) and calculate how much sulphur is required theoretically to prepare 20 litres of hydrogen sulphide at STP.

691. Indicate the main physical and chemical properties of hydrogen sulphide. What role does hydrogen sulphide play as it reacts with other substances, that of an oxidizing or of reducing agent? Justify your answer by considering the reaction of hydrogen sulphide with sulphuric acid and atmospheric oxygen from the standpoint of electronic theory.

692. Describe the oxidizing-reducing properties of sulphurous acid, and illustrate your answer with the equations of reactions of the acid with (a) iodine, (b) chloric acid. How many litres of sulphurous anhydride SO_2 at STP should be passed through a solution of chloric acid to reduce its 16.9 g to HCl?

693. What properties, oxidizing or reducing, does sulphurous acid show as it reacts with (a) magnesium, (b) hydrogen sulphide, (c) iodine? Which of its atoms or ions is responsible for these properties in each particular case?

694. What substance is commonly known as 'hypo'? Write its formula and the correct chemical name. Is it an oxidizing or reducing agent? Which atom in the molecule of the hypo is responsible for these properties? Indicate the method for preparation and the most important fields of application of the hypo.

695. What naturally occurring substances can be used in the manufacture of sulphuric acid by the contact and the chamber processes? What is the main difference between these two processes from the chemical point of view? Describe the importance of sulphuric acid.

696. What is the difference in the action of concentrated and dilute sulphuric acid on metals? Which atom or ion is the oxidant in the former and which in the latter case? Prove your answer by deriving the equations for the reactions (a) of concentrated sulphuric acid with magnesium bearing in mind the activity of magnesium; (b) of concentrated sulphuric acid with silver; (c) of dilute sulphuric acid with iron.

697. How many grams of sulphuric acid are required to dissolve 50 g of mercury? How many grams will be spent to oxidize it? Can dilute sulphuric acid be used to dissolve mercury?

698. Are the quantities of concentrated and dilute sulphuric acids required to dissolve 40 g of nickel equal? How many grams of H_2SO_4 will be spent to oxidize the nickel in each case? Which of the acid atoms or ions will be the oxidant in the former and in the latter cases?

699. Into what products can concentrated sulphuric acid be reduced in the reaction with metals? Indicate all possible cases of reduction of the acid, and derive the equations of the reactions with magnesium, zinc, mercury, paying special attention to the position occupied by these metals in the electromotive series.

700. Name the most important oxygen-containing sulphur acids. Describe their oxidation-reduction properties and indicate, which atom is responsible for its oxidizing or reducing properties. Give common names of salts of these acids.

701. Name the most important salts of sulphuric acid. Indicate their (a) chemical and industrial (if any) names; (b) the formula of a crystal hydrate; (c) practical uses.

702. Indicate the main chemical properties of hydrogen selenide and hydrogen telluride. Can hydrogen selenide be produced by the action of nitric acid on zinc selenide? Prove your answer by the equation of the reaction.

4. The Nitrogen Family

703. Give the general characteristics for the elements in the nitrogen family, indicating (a) their atomic structures; (b) the valence common for all elements in the group; (c) formulas of their compounds with hydrogen; (d) changes in their properties with the increasing atomic number and the cause of these changes (from the aspect of their atomic structures).

704. How many litres of ammonia at STP should be dissolved in one litre of water to obtain a 10 per cent solution? What are the main properties of ammonia solution? Do molecules of ammonia or ammonium ions prevail in the solution?

705. What equilibrium takes place in an aqueous solution of ammonia? Derive the ionic equation for the reaction of neutralization of ammonia solution with sulphuric acid. Which ions will remain in the solution?

706. How can ammonium salts be prepared? Give examples of reactions for their preparation. Name the specific properties of the ammonium salts. Indicate the most important salts and their practical uses.

707. The equivalent quantities of ammonium sulphate and table salt are mixed and heated. What is the resultant product? Write the equation for the reaction.

708. Derive the ionic equation for the reaction between an ammonium salt and an alkali. How many litres of ammonia at STP can be produced by the action of two litres of a $1.5N$ solution of an alkali on an ammonium salt?

709. Name the oxygen-containing acids of nitrogen. Characterize their oxidation-reduction properties. Derive the equations for the reactions between (a) concentrated nitric acid and mercury; (b) dilute nitric acid and calcium; (c) dilute nitric acid and silver.

710. Which is the most popular method for the manufacture of nitric acid at the present time? What naturally occurring substances are used in the manufacture of nitric acid by this method? Indicate the chemical reactions by which nitric acid is produced from the naturally occurring starting materials.

711. What nitrogen compounds are produced by directly bonding the atmospheric nitrogen? Write the reactions for their manufacture and the process conditions. Why has the discovery of the method for bonding the atmospheric nitrogen become so important for the humanity?

712. What is the principal difference in the action on metals of dilute nitric acid from that of hydrochloric and dilute sulphuric acids? Which atom or ion is the oxidant in the former and which in the latter case? Derive the equations for the reactions between dilute nitric acid and (a) mercury, (b) calcium, paying special attention to the position of these metals in the electromotive series.

713. What are the products of stepwise reduction of nitric acid? What is the greatest number of electrons that can be accepted by an atom of nitrogen $\overset{+V}{N}$ during reduction of nitric acid? Derive the equations for the reduction of nitric acid with zinc (a) to nitrous oxide; (b) to nitrogen; (c) to ammonia.

714. Derive the equations for the following reactions illustrating the oxidizing properties of nitric acid: (a) the action of concentrated nitric acid on sulphur; (b) the action of dilute nitric acid on lead; (c) the action of dilute nitric acid on magnesium. On what does the extent to which nitric acid is reduced in the reaction with various metals depend?

715. A lump of silver should be dissolved in nitric acid. What acid, concentrated or diluted, will suit the purpose better? In which case will the consumption of nitric acid be less?

716. Nitric acid is decomposed with heating to liberate oxygen and nitrogen dioxide. Which atoms are oxidized and which are reduced in this reaction?

717. What is the brown gas that is evolved during the action of concentrated nitric acid on metals? Of what molecules does it consist? Why does its colour darken with increasing and lighten with lowering temperature? Will this gas obey Boyle's law when compressed? Derive the equations for the reactions which take place during the dissolution of this gas in water and in an alkali solution.

718. How can ammonium nitrate, a valuable fertilizer, be produced, if the atmospheric nitrogen and water are only available? Write the chemical reactions by which this fertilizer can be manufactured.

719. Name the most important nitrogen fertilizers. How can they be prepared? Derive the equations for all necessary reactions.

720. By what chemical reactions can calcium nitrate be obtained, if frequently occurring materials are only used?

721. Taken separately are equal volumes of nitric oxide and nitrogen peroxide at standard pressure and a temperature of 50° C. Both gases are compressed to 5 atm. Which gas occupies a lesser volume now? Why?

722. Give examples of the reactions in which elementary nitrogen plays the role of the oxidizing agent and in which it poses as the reducing agent.

723. Why can nitric acid be used to prepare carbon dioxide from sodium carbonate, and cannot be used to prepare sulphurous anhydride from sodium sulphite? Illustrate your answer by equations of the reactions.

724. Explain the fact, why phosphorus occurs only in the form of its compounds, whereas nitrogen, its analogue, occurs in the free state. How can free phosphorus be obtained? Give the equation of the reaction and indicate, which step of the process is the oxidation-reduction reaction. What is oxidized and what is reduced?

725. Name two main allotropic modifications of phosphorus. What is the difference in their properties? Is this difference retained as they are converted into the gaseous state?

726. Prove that white and red phosphorus are only allotropic modifications of one element.

727. How can orthophosphoric acid be produced from (a) free phosphorus; (b) calcium phosphate? How much phosphorus and calcium phosphate are required to prepare 250 g of orthophosphoric acid? Can both processes be called oxidation-reduction ones?

728. Give names and formulas of ammonium salts of orthophosphoric acid. Why can ammonia be obtained by simply heating them, whereas in order to produce ammonia from ammonium chloride, the latter should be preliminarily mixed with lime or sodium hydroxide and the like?

729. To a solution containing one mole of H_3PO_4 one mole of Na_2HPO_4 was added and the solution was evaporated to dry. What is the product that remains after evaporation?

730. Name the most important natural compounds of phosphorus. Why are they almost not used directly as fertilizers, but require preliminary chemical treatment? What is this treatment? What are the superphosphate and precipitate? Write equations for the reactions by which they are produced.

731. Give the general characteristic of the elements of the arsenic subgroup, indicating (a) their atomic structures; (b) valence in compounds; (c) formulas and the properties of the oxides and hydroxides (separately for each element). How does the strengthening of the metal properties manifest itself in the transition from arsenic to antimony and bismuth?

732. Give the names and formulas for antimony hydroxides. What are their properties? To what does their solubility in alkalis indicate? In which of the following compounds does antimony play the role of a nonmetal and in which the role of a metal?

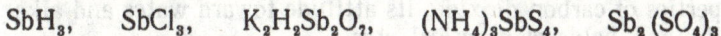

SbH_3, $SbCl_3$, $K_2H_2Sb_2O_7$, $(NH_4)_3SbS_4$, $Sb_2(SO_4)_3$

733. What substances are produced in the reaction between concentrated nitric acid and arsenic trisulphide? Derive the equation for the reaction.

734. What arsenic compound will be produced by reacting dilute sulphuric acid and zinc with arsenous acid anhydride As_2O_3? Derive the equation for the reaction and consider it from the oxidation-reduction point of view.

735. Why does a solution of antimony trichloride $SbCl_3$ become turbid when diluted with water? How can transparency be restored without filtering? Derive the molecular and ionic equations for both reactions.

736. If a zinc plate is immersed into a solution of antimony trichloride $SbCl_3$ acidified with hydrochloric acid, metallic antimony is deposited on the plate. What antimony compound will also be produced in these conditions? Express the processes by the appropriate equations.

737. What compounds are known as thioacids? How do they differ from common oxygen-containing acids? Write the ionic equations for the reactions by which ammonium salts of thioarsenous and thioantimonic acids are produced. Can free thioacids be obtained from them by the usual (what namely) method? Prove your answer by the equations of the reactions.

738. Bismuth easily dissolves in nitric acid but does not in hydrochloric and dilute sulphuric acids. What conclusion can be drawn as to the position of bismuth in the electromotive series? Derive the equation for the reaction which takes place in the former case. Can bismuth nitrate be purified by its recrystallization from an aqueous solution? Justify your answer.

5. Carbon and Silicon

739. Describe briefly the properties of carbon, indicating (a) its occurrence in nature; (b) allotropic modifications and the cause of the differences in their properties; (c) atomic structure of carbon and its valence in compounds; (d) oxides of carbon and their properties.

740. How can carbon dioxide be obtained in a laboratory? in industry? What naturally occurring materials can serve as the source of carbon dioxide? Describe the physical properties of carbon dioxide, its attitude toward water and alkalis, and fields of practical uses.

741. Why is hydrochloric acid (and not sulphuric acid) used to obtain carbon dioxide from marble? How many litres of carbon dioxide at STP can be prepared from 1 kilogram of marble containing 96 per cent of $CaCO_3$?

742. Describe briefly the properties of carbonic acid (methods for preparation, attitude toward heating, dissociation in an aqueous solution). Can a normal solution of carbonic acid be prepared? Can a neutral solution be obtained by adding an accurately equivalent quantity of an alkali to a solution of carbonic acid? Justify your answer.

743. Name the most important salts of carbonic acid. How can soda be obtained, if metallic sodium, hydrochloric acid, marble and water are available? Derive the equations for the reactions. What are the main industrial uses of soda?

744. In order to prepare soda, a solution of sodium hydroxide was divided into two equal portions. One of them was

saturated with carbon dioxide and then mixed with the
other portion. What was the product of saturation of the
first portion? What was the reaction which took place on
mixing both portions? Derive the equations for both reactions.

745. What chemical reaction should be used to purify carbon dioxide from an admixture of sulphurous anhydride?

746. How is soda prepared by the Solvay, or ammonium,
process? The starting products do not occur in nature; how
are they produced? What properties of sodium bicarbonate
are utilized during its separation from the by-product ammonium chloride?

747. What substance is known as potash? How is it prepared and what are its main industrial uses? How can potash
be prepared if potassium sulphate, barium hydroxide, calcium
carbonate and hydrogen chloride are available? Write the equations for the reactions.

748. A concentrated solution of soda was saturated with
carbon dioxide. The precipitated crystals were separated on
a filter and calcined. What is the final product? Derive the
equations for the reactions.

749. How can sodium carbonate be converted into sodium
bicarbonate? How much (in per cent) will the quantity of pure
sodium bicarbonate be reduced on calcining? How many
litres of carbon dioxide at STP can be obtained from 210 g
of sodium bicarbonate (a) by calcining; (b) by the action of
an acid?

750. Describe the main chemical properties of carbon monoxide. To what class of oxides does it belong? How can
it be obtained in the pure form? Why can carbon monoxide
burn to turn into carbon dioxide, whereas water vapour cannot burn, although there is a compound richer in oxygen
than water, namely, hydrogen peroxide?

751. In what cases is carbon monoxide produced in burning? Why is the danger of poisoning reduced (with the chimney closed) with the lowering temperature of coals? Name the
most important types of gaseous fuel which contain carbon
monoxide.

752. Knowing the formation heats of carbon dioxide (94
kcal), carbon monoxide (26.4 kcal) and water vapour (57.8
kcal), prove that the process of formation of producer gas is

accompanied by the liberation of heat, whereas the process of production of water gas by the absorption of heat.

753. How much heat will be liberated on burning one cubic metre of lighting gas containing 50 per cent of hydrogen, 35 per cent of methane, 8 per cent of carbon monoxide, 2 per cent of ethylene and 5 per cent of noncombustible admixtures?

754. Describe briefly silicon, indicating (a) its atomic structure and possible valences; (b) chemical properties; (c) the formula of its simplest compound with hydrogen; (d) properties of silica, its attitude toward water, acids and alkalis.

755. In what forms does silicon occur in nature? Name its most important natural modifications. What silicate is the most widely spread in nature? Write in the form of oxides the formulas of the following silicates: $H_2Mg_3Si_4O_{12}$ (talcum); $Na_2Al_2Si_6O_{16}$ (albite); $CaAl_2Si_2O_3$ (anorthite).

756. What are the main natural silicates that are used as raw materials in the silicate industry? How are glass and cement produced? What is carborundum, how is it prepared and for what purposes used?

757. What silicates can be prepared from chalk, sand, soda and potash? Give their names and formulas, and derive the equations for the reactions by which they are produced from these materials.

758. How is quartz glass produced? What are its advantages over common glass? What acid can be stored neither in common glass nor in quartz glass vessels? Why?

6. Metals of the First Group of the Periodic System

759. Describe briefly the properties of alkali metals, indicating (a) their position in the Periodic System and the atomic structure; (b) specific physical properties; (c) basic chemical properties; (d) attitude toward water and atmospheric oxygen. Name oxides and hydroxides formed by the alkali metals, and indicate their properties.

760. Why are potassium and sodium called alkali metals? What property of these metals is responsible for this name?

Name the most important natural compounds of potassium and sodium and indicate their practical uses.

761. What is the difference between the electrolytic methods for preparation of the alkali metals and caustic alkalis? What chemical reactions take place in the former and the latter cases? Express them by equations.

Indicate the most important uses of metallic sodium and sodium hydroxide.

762. The alkali metals are very active reducers. How does this property relate to their atomic structure? How is the strong reducing power manifested during reaction of the alkali metals with dilute nitric acid? With hydrogen? Illustrate your answer by reaction equations.

763. Indicate two industrial methods for preparing sodium hydroxide. What are the by-products of these processes? Express by the equations all chemical processes which take place during the preparation of sodium hydroxide by both methods.

764. How can potassium hydroxide, potassium hypochlorite and potassium chlorate be electrolyzed from a solution of potassium chloride? Write the equations of the reactions which take place.

765. Why should caustic alkalis be stored in stoppered vessels? If otherwise, into what substances can the alkalis be converted?

766. What substances are produced during saturation of sodium hydroxide solutions with (a) chlorine; (b) sulphur dioxide; (c) carbon dioxide; (d) hydrogen sulphide? Derive the equations of these reactions.

767. Commercial sodium hydroxide often contains considerable quantities of soda. How can it be detected most easily? How can sodium hydroxide solution be freed from the soda admixture? Derive the equations of the appropriate reactions.

768. Storage of a sodium hydroxide solution in a glass container erodes appreciably the glass. Explain this phenomenon and derive the equation for the reaction.

769. What is the essential difference between the reaction of the alkali metals with hydrogen and that of the halogens with hydrogen? Exemplify both types of the reactions and consider them from the oxidation-reduction viewpoint.

770. During neutralization of strong acids, soda is often used instead of sodium hydroxide. Calculate, how many grams of soda are required to replace 100 g of sodium hydroxide. What other salts of sodium can be used instead of sodium hydroxide, if hydrogen ion bonding is only meant?

771. Sodium hydride reacts with water to liberate hydrogen and to form sodium hydroxide. Derive the equation for the reaction. Indicate, which ions are oxidized and which are reduced, and determine the thermal effect of the reaction, knowing that the formation heat of the hydride is 13.1 kcal/mole, and of sodium hydroxide 102 kcal/mole.

772. Indicate the difference in chemical properties of the alkali metals and those of the copper subgroup (attitude toward water, air, dilute acids; the strength of the formed bases; reducing properties). How can this difference be explained?

773. Write down the formulas of the oxide, chloride and sulphide of the monovalent and bivalent copper. How can the chloride of the monovalent copper be prepared from the chloride of the bivalent copper? Derive the equation of the reaction and indicate the reducing and the oxidizing agents.

774. How is copper refined electrolytically? Although iron, nickel, lead and tin contained in crude copper are transferred into solution they are not liberated at the cathode. Explain this phenomenon and give the equations for the reactions that take place at the cathode and the anode. Calculate, how much time is required to prepare 1 kilogram of pure copper at a current intensity of 200 A.

775. Almost all water-insoluble compounds of copper are easily dissolved in an aqueous solution of ammonia and in solutions of cyanates of the alkali metals owing to the formation of complexes. Express by the ionic equations the reactions of dissolution of $CuCl$ and $Cu_3(PO_4)_2$ in ammonia and in a solution of potassium cyanate, bearing in mind that the coordination number of the univalent copper is 2, and of the bivalent copper 4.

776. If an ammonia solution of copper hydroxide is strongly diluted with water, copper hydroxide precipitates. Explain this phenomenon taking into consideration the equilibria that are attained in the solution.

777. How can the fact be explained, that silver chloride is not precipitated in the action of sodium chloride on a solution of the complex salt $K[Ag(CN)_2]$ whereas the reaction between sodium sulphide and the same solution produces a precipitate of silver sulphide? Express the latter reaction by the ionic equation.

778. To determine silver content of a coin, its sample weighing 0.3 g was dissolved in nitric acid and the silver precipitated from the solution with hydrochloric acid. After washing and drying, the precipitate weighed 0.119 g. Calculate the percent content of silver in the coin.

7. Metals of the Second Group
of the Periodic System

779. Characterize briefly the alkali-earth metals, indicating (a) their atomic structure and the valence in various compounds; (b) the chemical activity, the attitude toward water and air; (c) the properties of oxides and hydroxides formed with these metals. How did the name alkali-earth metals' originate?

Can the ions of the alkali-earth metals play the role of reducing agents? Justify your answer.

780. Into what product does calcium turn on burning? Why are considerable quantities of heat liberated and is the smell of ammonia felt as the calcium, burnt in air, is wettened with water? What chemical reactions take place during this process?

781. How can quicklime be prepared? What is the essence of the process of lime slaking? Express the reactions by the equations and calculate how much heat will be evolved during slaking of 1 kilogram of lime, knowing that the formation heat of the quicklime is 236 kcal/mole, and of calcium oxide CaO 151.7 kcal/mole.

782. Why, when stored in the open, does lime lose its properties? Express the reaction by the equation.

783. Which compounds of the second group metals are widely used as mortars in building? How are they prepared, and what is responsible for their binding properties? Illustrate your answer by the equations of the appropriate reactions.

784. When 5 g of lime containing $CaCO_3$ admixture were dissolved in an acid, 140 ml of a gas were liberated (as measured at STP). What is the percent content of calcium carbonate in the lime?

785. Calcium hydride burns in the air, and when reacted with water, it liberates hydrogen. Write the equations for the reactions that take place in both cases and indicate what is oxidized and what is reduced in both cases.

786. How is metallic magnesium prepared and what are its main uses? Which metals in the second group cannot be electrolyzed from aqueous solutions of their salts? Why? Indicate all chemical processes that will take place during the electrolysis of an aqueous solution of magnesium chloride.

787. Magnesium hydroxide is soluble in a solution of ammonium chloride and is not in solutions of sodium chloride or potassium chloride. Recall the requisite conditions for dissolution of precipitates and explain this phenomenon.

788. Characterize briefly the elements of the zinc subgroup, indicating (a) their atomic structure; (b) the valence in various compounds; (c) the activity, compared to the alkali earth metals; (d) oxides and hydroxides and their properties; (e) the attitude toward water, acids and alkalis.

In which metal of the zinc subgroup are the metallic properties the least of all marked? How is it manifested?

789. What is the industrial method for preparing zinc from natural zinc sulphide? Write the equations for all the chemical processes that take place during the preparation of metallic zinc.

790. When hydrogen sulphide is passed through a solution of zinc chloride, only insignificant precipitate of zinc sulphide falls out, since owing to its solubility in hydrochloric acid the reaction is not complete. If sodium acetate is added preliminarily to the solution, zinc is precipitated completely with hydrogen sulphide in the form of zinc sulphide. Explain this phenomenon.

791. Metallic mercury often contains admixtures of zinc, tin and lead. To purify mercury, it is shaken with a saturated solution of mercuric sulphate. What principle underlies this method of mercury purification? Write down the equations of the reactions.

792. With an aqueous solution of which substance (from those given below) will mercury react?

HCl, HNO_3, $NaOH$, $AgNO_3$, $CuCl_2$, $ZnCl_2$

Write the equations for the reactions.

793. Substances are formed from zinc chloride, cadmium chloride, or mercuric chloride and excess sodium hydroxide, to what class of chemical compounds will they relate? Write down the ionic equations for the reactions.

794. A piece of brass was dissolved in nitric acid. The solution was divided into two portions; excess ammonia was added to one portion and excess alkali to the other. Will thus formed compounds (what namely?) of copper and zinc be present in the solution or will they precipitate?

8. Hardness of Water

Hardness of natural water is due to the presence of salts of bivalent metals, mainly bicarbonates and sulphates of calcium, magnesium and iron. Hardness caused by the presence of bicarbonates is called *temporary* or *removable*, to distinguish it from *permanent* hardness which is due to the sulphates, chlorides and other salts. The sum of the temporary and permanent hardness is referred to as the *total hardness* of water.

The hardness of water is expressed in the USSR as the total number of milligram-equivalents of calcium and magnesium ions per litre of water *. The hardness of water of one milligram-equivalent indicates that one litre of the water contains one milligram-equivalent of calcium (or magnesium) ions, or one milligram-equivalent of $Ca(HCO_3)_2$ or $CaSO_4$ ($Mg(HCO_3)_2$ or $MgSO_4$).

Prior to 1952, the hardness of water used to be measured in degrees of hardness, showing the number of grams of calcium oxide present in 100 litres of water. The contents of magnesium and other bivalent metals were recalculated for an equivalent amount of CaO. One degree of hardness equals 0.35663 mg-equiv of calcium or magnesium ion.

One of the methods used for determining the hardness of water is titration of its certain volume with a standard solution of hydrochloric acid. Gravimetric methods for determination of the hardness are also used.

---

* It should be remembered that one milligram-equivalent (mg-equiv) is the quantity of a substance equal numerically to its equivalent weight and expressed in milligrams.

Softening of water consists in removal of calcium and magnesium ions from water.

Here are a few exemplary computations of the hardness of water and the quantities of reagents required to remove it.

Example 1. The quantity of a $0.12N$ solution of hydrochloric acid required to titrate 100 ml of water containing calcium bicarbonate is 2.5 ml. Calculate the hardness of the water.

Solution. From the conditions of the problem we can first determine the normality of the aqueous solution of the bicarbonate. Let the normality of the solution, that is the number of gram-equivalents of the bicarbonate in one litre of the water, be x, then

$$100 : 2.5 = 0.12 : x \quad \text{or} \quad 100x = 2.5 \times 0.12$$

whence

$$x = \frac{2.5 \times 0.12}{100} = 0.003 \text{ g-equiv}$$

It follows therefore that one litre of the water contains 3 mg-equiv of calcium bicarbonate or 3 mg-equiv of calcium ions.

The hardness of the water is 3 mg-equiv.

Example 2. How many grams of soda Na_2CO_3 should be added to 10 litres of water to remove its total hardness of 4.64 mg-equiv?

Solution. It follows from the equations of the reactions

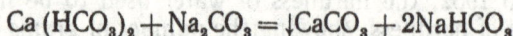

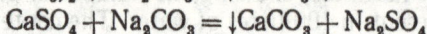

$$Ca(HCO_3)_2 + Na_2CO_3 = \downarrow CaCO_3 + 2NaHCO_3$$
$$CaSO_4 + Na_2CO_3 = \downarrow CaCO_3 + Na_2SO_4$$

that one gram-molecule of soda reacts with one gram-molecule of calcium bicarbonate or with one gram-molecule of calcium sulphate. By recalculating for milligram-equivalents, we can find that one mg-equiv of soda interacts with 1 mg-equiv of $Ca(HCO_3)_2$ or $CaSO_4$, that is with 1 mg-equiv of calcium ions.

Since the molecular weight of soda is 106, and its equivalent weight is half its molecular weight, 1 mg-equiv of soda weighs 53 mg. Ten litres of water contain $4.64 \times 10 = 46.4$ mg-equiv of calcium ions.

The quantity of soda (x), which is required to remove the hardness, can be found from the proportion:

$$x : 53 = 46.4 : 1$$

whence

$$x = 53 \times 46.4 = 2,459.2 \text{ mg or } 2.46 \text{ g}$$

Example 3. Calculate the permanent hardness of water, knowing that 10.8 g of anhydrous borax $Na_2B_4O_7$ were spent to remove calcium ions contained in 50 litres of the water.

Solution. When borax reacts with water containing calcium sulphate, calcium ions are precipitated:

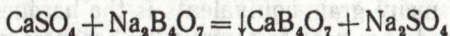

$$CaSO_4 + Na_2B_4O_7 = \downarrow CaB_4O_7 + Na_2SO_4$$

From the equation of the reaction it follows that to precipitate calcium ions in the form of CaB_4O_7, 1 g-equiv of borax should be taken per 1 g-equiv of calcium sulphate (that is per 1 g-equiv of calcium ions) or 1 mg-equiv of borax per 1 mg-equiv of calcium ions.

The molecular weight of borax is 202. It follows therefore that its 1 mg-equiv weighs 101 mg. In order to precipitate calcium ions contained in 50 litres of water 10.8 g or 10,800 mg of borax were spent. The number of milligram-equivalents of calcium ions contained in 50 litres of the water is found from the proportion:

$$x : 1 = 10,800 : 101$$

whence

$$x = \frac{10,800}{101} = 107 \text{ mg-equiv}$$

The found quantity of milligram-equivalents of calcium ions is contained in 50 litres of the water. Hence its hardness is

$$\frac{107}{50} = 2.14 \text{ mg-equiv}$$

PROBLEMS

795. The presence of what salts in natural water is responsible for its hardness? Why more soap is required if hard water is used for laundry? Will the presence of calcium chloride in water change the soap requirement?

796. Why is the hardness of water due to the presence of bicarbonates of calcium and magnesium called temporary or removable? What chemical reactions will take place (a) during boiling of hard water containing $Ca(HCO_3)_2$? (b) if soda is added to the water? (c) if sodium hydroxide is added to the water?

797. Both temporary and permanent hardness of water can be removed by adding soda. Can soda be replaced by (a) potassium carbonate? (b) barium hydroxide which is readily soluble in water? Prove your answers by the equations of the reactions.

798. How many gram-equivalents is the hardness of water if 15.9 g of soda are used to remove the hardness in 100 litres of water?

799. What ions, when introduced into natural water, can remove (a) its temporary hardness; (b) permanent hardness?

800. How many grams of quicklime should be added to 1000 litres of water to remove its temporary hardness of 2.86 mg-equiv?

801. Calculate temporary hardness of water knowing that 5 ml of a $0.1N$ solution of hydrochloric acid were spent to react with the bicarbonate contained in 100 ml of the water.

802. The hardness of water containing only calcium bicarbonate is 1.785 mg-equiv. Determine the quantity of the bicarbonate contained in one litre of the water.

803. What is the temporary hardness of water whose one litre contains 0.146 g of magnesium bicarbonate?

804. Calculate permanent hardness of water knowing that 10.8 g of anhydrous borax were spent to remove calcium ions contained in 50 litres of the water.

805. An analysis has shown that one litre of the test water contains 42 mg of magnesium ions and 112 mg of calcium ions. Calculate the total hardness of the water.

9. Metals of the Third and the Fourth Groups of the Periodic System

806. Characterize briefly boron, indicating (a) its position in the Periodic System, its atomic structure and the valence in various compounds; (b) properties of boron oxide; (c) the most important compounds of boron.

807. Why is borax used in soldering? Write the equations for the chemical reactions that take place during fusing borax with oxides of copper, iron and nickel (CuO, Fe_2O_3 and NiO).

808. Write the equations for the chemical reactions that take place when boric acid is heated. What salt will be produced on the interaction of a solution of $Ba(OH)_2$ with boric acid?

809. Characterize briefly aluminium, indicating (a) its atomic structure; (b) the valence in various compounds; (c) the attitude toward water, acids and alkalis; (d) the properties of its oxide and hydroxide. What facts indicate to the weak basic properties of aluminium hydroxide?

810. From what naturally occurring materials is aluminium produced? Write the equations for the reactions that take place at the anode and the cathode during preparation of aluminium. Indicate the main uses of aluminium.

811. To purify bauxite from Fe_2O_3 admixture it is fused with sodium hydroxide. The melt is then treated with water, and carbon dioxide is passed through the filtered solution. The precipitate is separated on a filter and calcined. Write the equations for the reactions that take place, and indicate at what stage the iron is separated from aluminium.

812. Aluminium stands far to the left of hydrogen in the electromotive series, but it does not displace hydrogen from water, and easily displaces it from an aqueous solution of an alkali. Why? What role does the alkali play here? Write the equations for all the reactions and calculate how much aluminium is spent for the production of a litre of hydrogen.

813. On the interaction between solutions of aluminium salts and sulphides of alkali metals, aluminium hydroxide is formed. Explain, why the product of this reaction is not aluminium sulphide and write the equation for the reaction.

814. On addition of a few drops of a solution of an aluminium salt to a solution of alkali, no precipitate falls out, but when several drops of an alkali solution are added to a solution of an aluminium salt, a precipitate is formed. Why? What is the composition of the precipitate? Write the equations for the reactions which take place in both cases.

815. How can (a) ammonia, (b) aluminium hydroxide, (c) barium sulphate, and (d) potassium aluminate be produced from aluminium ammonium sulphate? Write down the ionic equations of the reactions.

816. How can the extraordinary similarity of chemical properties of lanthanides be explained? All lanthanides are placed into one box of the Periodic Table. Can they be called isotopes on this ground? Justify your answer.

817. Name the metals of the fourth group whose atomic structure is similar to that of carbon. Indicate their oxides and hydroxides, and their properties. How do the nonmetal properties show in these elements? In which element are the nonmetal properties most manifest?

818. Indicate the most important properties of metallic tin, viz., its density, melting point, attitude toward air, water and acids. Why does iron rapidly corrode in places where the tin coat is damaged?

819. Derive the equations for the reactions according to the schemes:

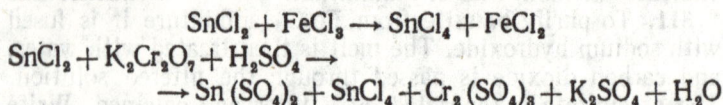

$$SnCl_2 + FeCl_3 \longrightarrow SnCl_4 + FeCl_2$$
$$SnCl_2 + K_2Cr_2O_7 + H_2SO_4 \longrightarrow$$
$$\longrightarrow Sn(SO_4)_2 + SnCl_4 + Cr_2(SO_4)_3 + K_2SO_4 + H_2O$$

and consider them from the oxidation-reduction viewpoint. What property of the bivalent tin ions is illustrated here?

820. How can sodium thiostannate be prepared from metallic tin? Derive the molecular and ionic equations for the necessary reactions. Sulphides of what elements can rapidly form thiosalts? How can a thioanhydride of the corresponding acid be prepared from a thiosalt?

821. Indicate the important physical constants (density and the melting point) of lead, its attitude toward air, water and acids. Why is lead insoluble in hydrochloric and dilute sulphuric acids although it stands to the left of hydrogen in the electromotive series? Indicate the main uses of lead.

822. An alloy of lead and tin was heated with concentrated nitric acid until the reaction was completed. The undissolved residue was separated on a filter, dried and calcined. What is the residue composition? What remained in the solution?

823. Can PbO_2 be called zinc peroxide? Justify your answer bearing in mind the properties of this oxide.

824. What salts can be produced by fusing lead dioxide with (a) calcium oxide? (b) sodium hydroxide? Give their names and write the equations for the reactions in which these salts are formed. Is lead a metal or a nonmetal here?

825. Why are lead oxides Pb_2O_3 and Pb_3O_4 called mixed? To what class of compounds would it be more correct to refer them? Justify your answer by deriving the equations of the reactions in which dilute nitric acid acts upon these oxides.

826. Bearing in mind the salt forming property of the oxides Pb_2O_3 and Pb_3O_4, and also knowing how lead dioxide reacts with hydrochloric acid, write the equations for the reactions between the two oxides and the acid.

10. Metals of the Sixth, Seventh and Eighth Groups of the Periodic System

827. Describe briefly the chemical properties of chromium, indicating (a) its position in the Periodic System and the atomic structure; (b) its attitude toward air, water and acids; (c) its oxides and hydroxides, and their properties. Why do the metal properties prevail in chromium, whereas sulphur, which stands in the same group, is a typical nonmetal?

828. What chromium compounds are characterized by the oxidizing properties? Give an example of the reaction where these properties are manifested. How does the valence of chromium change in these reactions?

829. Excess alkali solution was added to a solution of chromium sulphate, and bromine water was then added dropwise until the green colour of the liquid changed to yellow. Write the equations for all the reactions.

830. How can chrome alums be prepared from potassium dichromate? Derive the equations for the reactions that should be carried out. How many grams of potassium dichromate are required to prepare 1 kilogram of the alum?

831. All soluble compounds of the chromium oxide series are easily oxidized in alkali solution into salts of chromic acid (chromites are converted into chromates). Write the equations for the reactions, taking (a) chlorine and (b) hydrogen peroxide as the oxidizing agents.

832. When a solution of sodium chromite is boiled, or if ammonium chloride solution is added to it, chromium hydroxide precipitates. Why is the precipitate formed in both cases? Illustrate your answer by the equations of the reactions.

833. How can chromic anhydride be prepared from chromium chloride $CrCl_3$? Write the equations for the reactions.

834. As hydrogen sulphide is passed through a solution of potassium dichromate acidified with sulphuric acid, the orange colour of the solution changes gradually to green, and sulphur precipitates. Explain the change in the colour and write the equation of the reaction.

835. How many litres of chlorine will be liberated on the interaction of one gram-molecule of sodium dichromate with excess hydrochloric acid? What chromium compound will be formed in the reaction?

836. To what element in the third period is chromium similar in the compounds where it is (a) trivalent, (b) hexavalent? In what respects can they be compared?

837. How can tungsten and molybdenum be prepared from their compounds? What are the main uses of these metals?

838. Describe briefly the chemical characteristic of manganese, indicating (a) its position in the Periodic System and the atomic structure; (b) its oxides and hydroxides and their properties; (c) properties of the highest oxygen compounds of manganese. Give examples of salts where manganese has all its possible valences.

839. What manganese compound is most abundant in nature? How can it be produced artificially from (a) $MnCl_2$; (b) $KMnO_4$?

840. How can potassium permanganate be produced from manganese dioxide and potassium chlorate as the oxidizer? Write the equations for the reactions that will take place.

841. To solutions of the salts K_2CrO_4, K_2SO_4, K_2SeO_4 and K_2MnO_4 was added a certain amount of H_2SO_4. What solutions reacted with acid? Write the equations of the reactions.

842. What property is characteristic of permanganates? Write the equations for the reactions that take place between $KMnO_4$ and K_2S in a neutral solution and $KMnO_4$ and KI in an acid solution.

843. How many grams of $KMnO_4$ are required to oxidize 7.6 g of $FeSO_4$ in a neutral and in an acid solution?

844. To which group of the Periodic System does iron belong? Indicate its atomic structure, valence in various compounds, the attitude toward water and acids, name its oxides and hydroxides and their properties.

845. Which compounds of iron are more stable, ferric or ferrous? How can (a) a salt of the trivalent iron be converted into a salt of the divalent iron, and (b) a salt of the divalent iron into a salt of the trivalent iron? Exemplify the reactions.

846. During iron smelting out of magnetite ore, one of the chemical reactions that take place in the blast furnace is

$$Fe_3O_4 + CO \rightleftharpoons 3FeO + CO_2$$

Determine the thermal effect of this reaction, knowing that the formation heat of Fe_3O_4 is 266.5 kcal/mole, and that of FeO is 64.5 kcal/mole. Into what direction will the equilibrium be displaced with the increasing temperature?

847. What is the theoretical thermal effect (neglecting the heat loss due to radiation, heating of the crucible, etc.) in the reduction of ferric oxide with aluminium, if the mean specific heat of the reaction products is assumed to be 0.18 cal/g·deg? The formation heat of Fe_2O_3 is 195 kcal/mole and of Al_2O_3, 393 kcal/mole.

848. What facts indicate that ferric hydroxide is a weak base? As soda reacts with a solution of ferric chloride $FeCl_3$, ferric hydroxide, and not ferric carbonate is produced. Explain this phenomenon and derive the equations for the reaction presenting it in two steps.

849. Sodium hydroxide can be obtained by calcining anhydrous soda with ferric oxide. After treating the melt with hot water, sodium hydroxide (which transfers into solution) and ferric oxide are formed. Bearing in mind that Ee_2O_3 has weak acid properties, explain this method and write the equations for the reactions by which sodium hydroxide is produced.

850. To determine $FeSO_4$ content of a green vitriol solution, its ten millilitres were placed into a beaker, acidified with sulphuric acid, and then $0.02M$ solution of potassium permanganate was added in drops until a pink colour developed from a single drop. What does the pink colour mean? How many moles of $FeSO_4$ does one litre of the test solution contain, if 30 ml of the permanganate solution were spent to develop the pink colour?

851. What ions are contained in a solution of $Ba_2[Fe(CN)_6]$? Which of the substances will react with this solution:

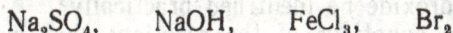

$$Na_2SO_4, \qquad NaOH, \qquad FeCl_3, \qquad Br_2$$

Write the ionic equations for the reactions that will take place.

852. How can Berlin blue be prepared from green vitriol, nitric acid and potassium cyanate? Write the equations of the corresponding reactions.

853. If a fabric dyed with Berlin blue is stained with alkali it turns brown. When treated with dilute hydrochloric acid the brown spot will disappear. Explain this phenomenon and write the equations of the reactions.

854. A solution of cobaltous sulphate $CoSO_4$ reacts with an alkali on heating to yield a dirty pink precipitate, and a solution of nickelous sulphate $NiSO_4$ yields a pale green precipitate. When both precipitates are treated with bromine the colour turns black. Write the equations of the reactions.

855. Derive the equations for the reactions that take place between sulphuric acid and nickelic hydroxide $Ni(OH)_3$ and between hydrochloric acid and cobaltic oxide Co_2O_3. What other oxides in your knowledge react similarly with hydrochloric acid?

11. Supplementary Problems

856. Such substances as abrasive paper, gypsum, soda, saltpetre, potash are well known to everyone. Give their chemical names, formulas, and indicate their practical uses.

857. What are the chemical formulas of metal compounds used as dyes, such as white zinc, mummy, white lead, chrome yellow, Berlin blue, Paris green, chrome green, cinnabar, minium?

858. What are the main uses of vanadium, molybdenum, and tungsten (in the free state or in compounds)?

859. Write the formulas and indicate the most important uses of soluble glass, carborundum, chlorinated lime, sodium alum, corrosive sublimate.

860. Which of the abundant metals form acid oxides? Write the names and the formulas of the corresponding acids and give examples of their salts. What properties, oxidizing or reducing, do these acids possess?

861. Name hydroxides of the most important and abundant metals that possess amphoteric properties. How can an amphoteric hydroxide be identified practically?

Write the equations of the reactions that would prove amphoteric properties of stannous hydroxide.

862. Name the most important polyvalent metals. Indicate their possible valences, formulas of the salt-forming oxides and hydroxides and their properties. How does the character of oxides and hydroxides of a polyvalent metal change with increasing valence?

863. Name the most important natural compounds that can be used for preparing zinc, mercury, aluminium, tin, lead, chromium and manganese.

864. How do the nonmetal properties show themselves in tin, chromium and manganese, the elements that would be usually referred to metals? Indicate in which of the salts

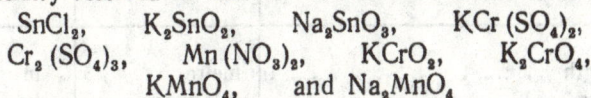

$$SnCl_2, \quad K_2SnO_2, \quad Na_2SnO_3, \quad KCr(SO_4)_2,$$
$$Cr_2(SO_4)_3, \quad Mn(NO_3)_2, \quad KCrO_2, \quad K_2CrO_4,$$
$$KMnO_4, \quad \text{and} \quad Na_2MnO_4$$

these elements have the properties of metals and in which of nonmetals.

865. Name all metal ions in your knowledge that can serve as reducing agents. Give examples of the reactions in which they act as reducing agents.

866. What gases can be obtained if sulphuric acid, sodium hydroxide, ammonium chloride, sodium sulphate and manganese dioxide are available?

Derive the equations for the corresponding reactions.

867. There are solutions of sodium hydroxide, sodium sulphite and potassium dichromate. Which of these solutions can absorb SO_2, CO_2, Cl_2 and H_2S? Write the equations for the reactions.

868. Give examples of reactions in which hydrogen and nitrogen molecules serve as oxidizers, and in which they are reducing agents?

869. Which of the hydroxides from the list below are soluble in alkalis?

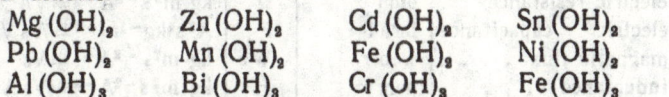

$$Mg(OH)_2 \quad Zn(OH)_2 \quad Cd(OH)_2 \quad Sn(OH)_2$$
$$Pb(OH)_2 \quad Mn(OH)_2 \quad Fe(OH)_2 \quad Ni(OH)_2$$
$$Al(OH)_3 \quad Bi(OH)_3 \quad Cr(OH)_3 \quad Fe(OH)_3$$

Express the reactions by the equations.

870. Is it possible to prepare a solution that would contain simultaneously Sn^{2+} and Hg^{2+}, Sn^{2+} and Fe^{2+}, Sn^{2+} and Fe^{3+}, SO_3^{2-} and MnO_4^-, $Cr_2O_7^{2-}$ and SO_4^{2-}?

Indicate which combinations are impossible, and why.

APPENDICES

1. The International System of Units

Basic SI units

Quantity	Unit	Symbol
length	metre	m
mass	kilogram	kg
time	second	s
electric current	ampere	A
thermodynamic temperature	degree Kelvin	°K
luminous intensity	candela	cd

Derived SI units

Quantity	Unit	Symbol	Definition
energy	joule	J	$kg\ m^2\ s^{-2}$
force	newton	N	$kg\ m\ s^{-2} = J\ m^{-1}$
power	watt	W	$kg\ m^2\ s^{-3} = J\ s^{-1}$
electric charge	coulomb	C	As
electric potential difference . . .	volt	V	$kg\ m^2\ s^{-3}A^{-1} = JA^{-1}s^{-1}$
electric resistance . .	ohm	Ω	$kg\ m^2\ s^{-3}A^{-2} = VA^{-1}$
electric capacitance	farad	F	$A^2\ s^4 kg^{-1}m^{-2} = As\ V^{-1}$
magnetic flux	weber	Wb	$kg\ m^2 s^{-2}A^{-1} = Vs$
inductance	henry	H	$kg\ m^2 s^{-2}A^{-2} = Vs\ A^{-1}$
magnetic flux density	tesla	T	$kg\ s^{-2}A^{-1} = Vsm^{-2}$
luminous flux	lumen	lm	$cd\ sr$
illumination	lux	lx	$cd\ sr\ m^{-2}$
frequency	hertz	Hz	cycle per second
customary temperature, t	degree Celsius	°C	$t/°C = T/°K - 273.15$

Units allowed in conjunction with SI

Quantity	Unit	Symbol	Definition
length	parsec	pc	30.87×10^{15} m
area	barn	b	10^{-28} m^2
	hectare	ha	10^4 m^2
volume	litre	l	10^{-3} m$^3 =$ dm^3
pressure	bar	bar	10^5 N m^{-2}
mass	ton	t	10^3 kg $=$ Mg
kinematic viscosity, diffusion coefficient	stokes	St	10^{-4} m^2 s^{-1}
dynamic viscosity	poise	P	10^{-1} kg m^{-1} s^{-1}
magnetic flux density (magnetic induction)	gauss	G	10^{-4} T
radioactivity	curie	Ci	37×10^9 s^{-1}
energy	electronvolt	eV	$1.6021 = 10^{-19}$ J

Traditional units with SI equivalents

Physical quantity	Unit	Equivalent
length	ångström	10^{-10} m
	inch	0.0254 m
	foot	0.3048 m
	yard	0.9144 m
	mile	1.609 34 km
	nautical mile	1.853 18 km
area	square inch	645.16 mm^2
	square foot	0.092 903 m^2
	square yard	0.836 127 m^2
	square mile	2.589 99 km^2
volume	cubic inch	$1.638 71 \times 10^{-5}$ m^3
	cubic foot	0.028 316 8 m^3
	U. K. gallon	0.004 546 092 m^3
mass	pound	0.453 592 37 kg
	slug	14.593 9 kg

(Continued)

Physical quantity	Unit	Equivalent
density	pound/cubic inch	$2.767\,99 \times 10^4$ kg m^{-3}
	pound/cubic foot	16.0185 kg m^{-3}
force	dyne	10^{-5} N
	poundal	$0.138\,255$ N
	pound-force	$4.448\,22$ N
	kilogram-force	$9.806\,65$ N
pressure	atmosphere	101.325 kN m^{-2}
	torr	133.322 N m^{-2}
	pound (f)/sq in.	$6\,894.76$ N m^{-2}
energy	erg	10^{-7} J
	calorie (I. T.)	4.1868 J
	calorie (15°C)	4.1855 J
	calorie (thermochemical)	4.184 J
	BThU	$1\,055.06$ J
	foot poundal	$0.042\,140\,1$ J
	foot pound (f)	$1.355\,82$ J
power	horsepower	745.700 W
temperature	degree Rankine	$\dfrac{5}{9}$ °K
	degree Fahrenheit	$t/°\mathrm{F} = \dfrac{9}{5}\,T/°\mathrm{C} + 32$

2. Varour Pressure of Water (mm Hg)

Degrees C	Pressure	Degrees C	Pressure	Degrees C	Pressure
0	4.578	18	15.48	40	55.32
1	4.926	19	16.48	45	71.88
3	5.685	20	17.54	50	92.51
5	6.543	21	18.65	55	118.0
7	7.513	22	19.83	60	149.4
9	8.609	23	21.07	65	187.5
10	9.209	24	22.38	70	233.7
11	9.84	25	23.76	75	289.1
12	10.52	26	25.21	80	355.1
13	11.23	27	26.74	85	433.6
14	11.99	28	28.35	90	525.76
15	12.79	29	30.04	95	633.90
16	13.63	30	31.82	100	760.00
17	14.53	35	42.18		

3. Density of Aqueous Solutions of Some Alkalis at 20° C (g/cu cm)

%	KOH	NaOH	NH₃	%	KOH	NaOH
4	1.035	1.043	0.981	34	1.330	1.370
6	1.053	1.065	0.973	36	1.359	1.390
8	1.072	1.087	0.965	38	1.373	1.410
10	1.090	1.109	0.958	40	1.396	1.430
12	1.109	1.131	0.950	42	1.418	1.449
14	1.128	1.153	0.943	44	1.441	1.469
16	1.147	1.175	0.935	46	1.464	1.487
18	1.167	1.197	0.930	48	1.487	1.507
20	1.186	1.219	0.923	50	1.510	1.525
22	1.206	1.241	0.916			
24	1.226	1.263	0.910			
26	1.247	1.285	0.904			
28	1.267	1.306	0.898			
30	1.288	1.329	0.892			
32	1.309	1.349	—			

4. Density of Aqueous Solutions of Some Acids
at 20° C (g/cu cm)

%	H_2SO_4	HNO_3	HCl	%	H_2SO_4	HNO_3
4	1.025	1.020	1.018	54	1.435	1.334
6	1.038	1.031	1.028	56	1.456	1.345
8	1.052	1.043	1.038	58	1.477	1.356
10	1.066	1.054	1.047	60	1.498	1.367
12	1.080	1.066	1.057	62	1.520	1.377
14	1.095	1.078	1.068	64	1.542	1.387
16	1.109	1.090	1.078	66	1.565	1.396
18	1.124	1.103	1.088	68	1.587	1.405
20	1.139	1.115	1.098	70	1.611	1.413
22	1.155	1.128	1.108	72	1.634	1.422
24	1.170	1 140	1.119	74	1.657	1.430
26	1.186	1.153	1.129	76	1.681	1.438
28	1.202	1.167	1.139	78	1.704	1.445
30	1.219	1.180	1.149	80	1.727	1.452
32	1.235	1.193	1.159	82	1.749	1.459
34	1.252	1.207	1.169	84	1.769	1.466
36	1.268	1.221	1.179	86	1.787	1.472
38	1.286	1.234	1.189	88	1.802	1.477
40	1.303	1.246	1.198	90	1.814	1.483
42	1.321	1.259	—	92	1.824	1.487
44	1.338	1.272	—	94	1.831	1.491
46	1.357	1.285	—	96	1.836	1.495
48	1.376	1.298	—	98	1.837	1.501
50	1.395	1.310	—	100	1.831	1.513
52	1.415	1.322	—			

While solving problems, use the rounded atomic weights given below, unless otherwise specified in the condition of the problem:

Element	Symbol	At. wt.	Element	Symbol	At. wt.
Aluminium	Al	27	Lead	Pb	207
Antimony	Sb	122	Magnesium	Mg	24
Argon	Ar	40	Manganese	Mn	55
Arsenic	As	74	Mercury	Hg	201
Barium	Ba	137	Molybdenum	Mo	96
Bismuth	Bi	209	Neon	Ne	20
Boron	B	11	Nickel	Ni	59
Bromine	Br	80	Nitrogen	N	14
Cadmium	Cd	112	Oxygen	O	16
Calcium	Ca	40	Phosphorus	P	31
Carbon	C	12	Platinum	Pt	195
Chlorine	Cl	35.5	Potassium	K	39
Chromium	Cr	52	Silicon	Si	28
Copper	Cu	64	Silver	Ag	108
Fluorine	F	19	Sodium	Na	23
Gold	Au	197	Sulphur	S	32
Helium	He	4	Tin	Sn	119
Hydrogen	H	1	Tungsten	W	184
Iodine	I	127	Vanadium	V	51
Iron	Fe	56	Zinc	Zn	65

Common

N	0	1	2	3	4	5	6
10	0000	0043	0086	0128	0170	0212	0253
11	0414	0453	0492	0531	0569	0607	0645
12	0792	0828	0864	0899	0934	0969	1004
13	1139	1173	1206	1239	1271	1303	1335
14	1461	1492	1523	1553	1584	1614	1644
15	1761	1790	1818	1847	1875	1903	1931
16	2041	2068	2095	2122	2148	2175	2201
17	2304	2330	2355	2380	2405	2430	2455
18	2553	2577	2601	2625	2648	2672	2695
19	2788	2810	2833	2856	2878	2900	2923
20	3010	3032	3054	3075	3096	3118	3139
21	3222	3243	3263	3284	3304	3324	3345
22	3424	3444	3464	3483	3502	3522	3541
23	3617	3636	3655	3674	3692	3711	3729
24	3802	3820	3838	3856	2874	3892	3909
25	3979	3997	4014	4031	4048	4065	4082
26	4150	4166	4183	4200	4216	4232	4249
27	4314	4330	4346	4362	4378	4393	4409
28	4472	4487	4502	4518	4533	4548	4564
29	4624	4639	4654	4669	4683	4698	4713
30	4771	4786	4800	4814	4829	4843	4857
31	4914	4928	4942	4955	4969	4983	4997
32	5051	5065	5079	5092	5105	5119	5132
33	5185	5198	5211	5224	5237	5250	5263
34	5315	5328	5340	5253	5366	5378	5391
35	5441	5453	5465	5478	5490	5502	5514
36	5563	5575	5587	5599	5611	5623	5635
37	5682	5694	5705	5717	5729	5740	5752
38	5798	5809	5821	5832	5843	5855	5866
39	5911	5922	5933	5944	5955	5966	5977

Logarithms

7	8	9	Proportional parts								
			1	2	3	4	5	6	7	8	9
0294	0334	0374	4	8	12	17	21	25	29	33	37
0682	0719	0755	4	8	11	15	19	23	26	30	34
1038	1072	1106	3	7	10	14	17	21	25	28	31
1367	1399	1430	3	6	10	13	16	19	23	26	29
1673	1703	1732	3	6	9	12	15	18	21	24	27
1959	1987	2014	3	6	8	11	14	17	20	22	25
2227	2253	2279	3	5	8	11	13	16	18	21	24
2480	2504	2529	2	5	7	10	12	15	17	20	22
2718	2742	2765	2	5	7	9	12	14	16	19	21
2945	2967	2989	2	4	7	9	11	13	16	18	20
3160	3181	3201	2	4	6	8	11	13	15	17	19
3365	3385	3404	2	4	6	8	10	12	14	16	18
3560	3579	3598	2	4	6	8	10	12	14	15	17
3747	3766	3784	2	4	6	7	9	11	12	14	17
3927	3945	3962	2	4	5	7	9	11	12	14	17
4099	4116	4133	2	3	5	7	9	10	12	14	15
4265	4281	4298	2	3	5	7	8	10	11	13	15
4425	4440	4456	2	3	5	6	8	9	11	13	14
4579	4594	4609	2	3	5	6	8	9	11	12	14
4728	4742	4757	1	3	4	6	7	9	10	12	13
4871	4886	4900	1	3	4	6	7	9	10	11	13
5011	5024	5038	1	3	4	6	7	8	10	11	12
5145	5159	5172	1	3	4	5	7	8	9	11	12
5276	5289	5302	1	3	4	5	6	8	9	10	12
5403	5416	5428	1	3	4	5	6	8	9	10	11
5527	5539	5551	1	2	4	5	6	7	9	10	11
5647	5658	5670	1	2	4	5	6	7	8	10	11
5763	5775	5786	1	2	3	5	6	7	8	9	10
5877	5888	5899	1	2	3	5	6	7	8	9	10
5988	5999	6010	1	2	3	4	5	7	8	9	10

N	0	1	2	3	4	5	6
40	6021	6031	6042	6053	6064	6075	6085
41	6128	6138	6149	6160	6170	6180	6191
42	6232	6243	6253	6263	6274	6284	6294
43	6335	6345	6255	6365	6375	6385	6395
44	6435	6444	6454	6464	6474	6484	6493
45	6532	6542	6551	6561	6571	6580	6590
46	6628	6637	6646	6656	6665	6675	6684
47	6721	6730	6739	6749	6758	6767	6776
48	6812	6821	6830	6839	6848	6857	6866
49	6902	6911	6920	6928	6937	6946	6955
50	6990	6998	7007	7016	7024	7033	7042
51	7076	7084	7093	7101	7110	71·18	7126
52	7160	7168	7177	7185	7193	7202	7210
53	7243	7251	7259	7267	7275	7284	7292
54	7324	7332	7340	7348	7356	7364	7372
55	7404·	7412	7419	7427	7435	7443	7451
56	7482	7490	7497	7505	7513	7520	7528
57	7559	7566	7574	7582	7589	7597	7604
58	7634	7642	7649	7657	7664	7672	7679
59	7709	7716	7723	7731	7738	7745	7752
60	7782	7789	7796	7803	7810	7818	7825
61	7853	7860	7868	7875	7882	7889	7896
62	7924	7931	7938	7945	7952	7959	7966
63	7993	8000	8007	8014	8021	8028	8035
64	8062	8069	8075	8082	8089	8096	8102
65	8129	8136	8142	8149	8156	8162	8169
66	8195	8202	8209	8215	8222	8228	8235
67	8261	8267	8274	8280	8287	8293	8299
68	8325	8331	8338	8344	8351	8357	8363
69	8388	8395	8401	8407	8414	8420	8426

(Continued)

7	8	9	Proportional parts								
			1	2	3	4	5	6	7	8	9
6096	6107	6117	1	2	3	4	5	6	8	9	10
6201	6212	6222	1	2	3	4	5	6	7	8	9
6304	6314	6325	1	2	3	4	5	6	7	8	9
6405	6415	6425	1	2	3	4	5	6	7	8	9
6503	6513	6522	1	2	3	4	5	6	7	8	9
6599	6609	6618	1	2	3	4	5	6	7	8	9
6693	6702	6712	1	2	3	4	5	6	7	7	8
6785	6794	6803	1	2	3	4	5	5	6	7	8
6875	6884	6893	1	2	3	4	4	5	6	7	8
6964	6972	6981	1	2	3	4	4	5	6	7	8
7050	7059	7067	1	2	3	3	4	5	6	7	8
7135	7143	7152	1	2	3	3	4	5	6	7	8
7218	7226	7235	1	2	2	3	4	5	6	7	7
7300	7308	7316	1	2	2	3	4	5	6	6	7
7380	7388	7396	1	2	2	3	4	5	6	6	7
7459	7466	7474	1	2	2	3	4	5	5	6	7
7536	7543	7551	1	2	2	3	4	5	5	6	7
7612	7619	7627	1	2	2	3	4	5	5	6	7
7686	7694	7701	1	1	2	3	4	4	5	6	7
7760	7767	7774	1	1	2	3	4	4	5	6	7
7832	7839	7846	1	1	2	3	4	4	5	6	6
7903	7910	7917	1	1	2	3	4	4	5	6	6
7973	7980	7987	1	1	2	3	3	4	5	6	6
8041	8048	8055	1	1	2	3	3	4	5	5	6
8109	8116	8122	1	1	2	3	3	4	5	5	6
8176	8182	8189	1	1	2	3	3	4	5	5	6
8241	8248	8254	1	1	2	3	3	4	5	5	6
8306	8312	8319	1	1	2	3	3	4	5	5	6
8370	8376	8382	1	1	2	3	3	4	4	5	6
8432	8439	8445	1	1	2	2	3	4	4	5	6

N	0	1	2	3	4	5	6
70	8451	8457	8463	8470	8476	8482	8488
71	8513	8519	8525	8531	8537	8543	8549
72	8573	8579	8585	8591	8597	8603	8609
73	8633	8639	8645	8651	8657	8663	8669
74	8692	8698	8704	8710	8716	8722	8727
75	8751	8756	8762	8768	8774	8779	8785
76	8808	8814	8820	8825	8831	8837	8842
77	8865	8871	8876	8882	8887	8893	8899
78	8921	8927	8932	8938	8943	8949	8954
79	8976	8982	8987	8993	8998	9004	9009
80	9031	9036	9042	9047	9053	9058	9063
81	9085	9090	9096	9101	9106	9112	9117
82	9138	9143	9149	9154	9159	9165	9170
83	9191	9196	9201	9206	9212	9217	9222
84	9243	9248	9253	9258	9263	9269	9274
85	9294	9299	9304	9309	9315	9320	9325
86	9345	9350	9355	9360	9465	9370	9375
87	9395	9400	9405	9410	9415	9420	9425
88	9445	9450	9455	9460	9465	9469	9474
89	9494	9499	9504	9509	9513	9518	9523
90	9542	9547	9552	9557	9562	9566	9571
91	9590	9595	9600	9605	9609	9614	9619
92	9638	9643	9647	9652	9657	9661	9666
93	9685	9689	9694	9699	9703	9708	9713
94	9731	9736	9741	9745	9750	9754	9759
95	9777	9782	9786	9791	9795	9800	9805
96	9823	9827	9832	9836	9841	9845	9850
97	9868	9872	9877	9881	9886	9890	9894
98	9912	9917	9921	9926	9930	9934	9939
99	9956	9961	9965	9969	9974	9978	9983

| 7 | 8 | 9 | Proportional parts |||||||||
---	---	---	1	2	3	4	5	6	7	8	9
8494	8500	8506	1	1	2	2	3	4	4	5	6
8555	8561	8567	1	1	2	2	3	4	4	5	5
8615	8621	8627	1	1	2	2	3	4	4	5	5
8675	8681	8686	1	1	2	2	3	4	4	5	5
8733	8739	8745	1	1	2	2	3	4	4	5	5
8791	8797	8802	1	1	2	2	3	3	4	5	5
8848	8854	8859	1	1	2	2	3	3	4	5	5
8904	8910	8915	1	1	2	2	3	3	4	4	5
8960	8965	8971	1	1	2	2	3	3	4	4	5
9015	9020	9025	1	1	2	2	3	3	4	4	5
9069	9074	9079	1	1	2	2	3	3	4	4	5
9122	9128	9133	1	1	2	2	3	3	4	4	5
9175	9180	9186	1	1	2	2	3	3	4	4	5
9227	9232	9238	1	1	2	2	3	3	4	4	5
9279	9284	9289	1	1	2	2	3	3	4	4	5
9330	9335	9340	1	1	2	2	3	3	4	4	5
9380	9385	9390	1	1	2	2	3	3	4	4	5
9430	9435	9440	0	1	1	2	2	3	3	4	4
9479	9484	9489	0	1	1	2	2	3	3	4	4
9528	9533	9538	0	1	1	2	2	3	3	4	4
9576	9581	9586	0	1	1	2	2	3	3	4	4
9624	9628	9633	0	1	1	2	2	3	3	4	4
9671	9675	9680	0	1	1	2	2	3	3	4	4
9717	9722	9727	0	1	1	2	2	3	3	4	4
9763	9768	9773	0	1	1	2	2	3	3	4	4
9809	9814	9818	0	1	1	2	2	3	3	4	4
9854	9859	9863	0	1	1	2	2	3	3	4	4
9899	9903	9908	0	1	1	2	2	3	3	4	4
9943	9948	9952	0	1	1	2	2	3	3	4	4
9987	9991	9996	0	1	1	2	2	3	3	3	4

ANSWERS TO PROBLEMS

1. (Fe) : (S)=7 : 4; 63.63% Fe. **2.** (Ca) : (Br)=1 : 4; 20% Ca. **3.** (P) : : (O)=31 : 40; 43.66% P. **4.** (C) : (H)=3 : 1; 75% C. **5.** (C) : (S)=3 : 16; 84.2% S. **6.** 108 g. **7.** 96.15% As. **8.** (Cu) : (O)=4 : 1; (S) : (O)=1 : 1. **9.** 45 g. **10.** (Ca) : (C) : (O)=10 : 3 : 12. **11.** 9. **12.** 17.43. **13.** 12.14. **14.** 56.2. **15.** E_{Br}=79.91; E_M=9. **16.** 32.69. **17.** 1.76 g. **18.** 126.92. **19.** 10 litres. **20.** E_{ox}=24; E_M=16. **21.** 19. **22.** E_M=56.2; V_{H_2}=3.36 litres. **23.** 54.2. **24.** 45. **25.** 3 : 5. **26.** 1 : 2. **27.** 3 atm or 303,975 N/sq m. **28.** 2.4 atm or 243, 180 N/sq m. **29.** 839 ml. **30.** 1,800 litres. **31.** 22.4 litres. **32.** 746 ml. **33.** 273°C. **34.** 127°C. **35.** 797.2 mm Hg or 106, 280 N/sq m. **36.** 102.9 mm Hg or 13,720 N/sq m. **37.** —7°C. **38.** 115°C. **39.** 819 ml. **40.** 444 ml. **41.** 4.8 litres. **42.** 39.1. **43.** 200 mm Hg or 26,660 N/sq m, 600 mm Hg or 79,990 N/sq m. **44.** 159.6 mm Hg or 21,280 N/sq m. **45.** 6.825 litres. **46.** 750 mm Hg or 99,990 N/sq m. **47.** 0.4 litre. **48.** 34% and 66%. **49.** 624 mm Hg or 83,190 N/sq m. **50.** Total pressure is 687 mm Hg or 91,590 N/sq m. **51.** COCl$_2$. **52.** N$_2$O. **53.** C$_8$H$_8$. **54.** 4. **55.** (NH$_3$) : (Cl$_2$)=2 : 3. **57.** 2 : 5. **58.** 8 litres. **59.** 7.5 litres. **60.** 15 litres. **61.** 44% O$_2$ and 53% H$_2$. **62.** 40% NO$_2$ and 60% O$_2$. **63.** One volume of SO$_2$, 9 volumes of SO$_3$, 5.5 volumes of O$_2$. **64.** 0.15 atm or 15,200 N/sq m. **65.** Reduces 1/4 initial pressure. **66.** Will not change. **67.** Will reduce 1/7 initial pressure. **68.** Pressure did not change. Composition of the mixture: 60% Cl$_2$, 30% HCl and 10% H$_2$. **70.** 1,700 times as great. **71.** Oxygen 22.4 litres, glycerol about 73 ml, water 18 ml. **72.** 43.08 litres. **75.** 33.6 litres. **76.** 1 : 16 : 2. **77.** 5.6 cu m CO$_2$, 11.2 cu m H$_2$, 22.4 cu m CH$_4$. **78.** 24.04 litres. **79.** 44.6 moles. **80.** About one litre. **81.** 27×10^{18}. **82.** 0.5 g. **84.** 1.08 times as great in the latter case. **85.** 3.34×10^{22}. **86.** 7.1×10^9. **87.** 0.2 atm or 20,265 N/sq m. **88.** 1.12 atm or 113,480 N/sq m. **89.** 2.24 atm or 226,970 N/sq m. **90.** 0.541 atm (411 mm Hg) or 54,820 N/sq m. **91.** 70 atm or 7,092,750 N/sq m. **92.** 0.64 atm or 64,850 N/sq m, 1 atm or 101,325 N/sq m. **93.** 1.56 g. **94.** 1.295 kg. **95.** 1 g. **96.** 63.4 g. **97.** 0.214 kg. **98.** 1.54 g. **99.** 1.6 kg. **100.** 2.44 g. **101.** 2.52 g, D=1.17. **102.** 517 g. **103.** About 73 kg. **107.** 1.6 times as light. **110.** 0.517. **111.** 2.8 litres. **113.** 0.401. **114.** 28. **115.** 8. **116.** Mercury molecules are monoatomic. **117.** 4. **120.** 2.8; 80.9. **121.** 34. **122.** 58. **123.** 156. **124.** 71. **125.** 1 kg. **126.** 2.683 kg. **127.** 10 g. **128.** 820 litres. **129.** 39.6 litres. **130.** 936 mm Hg or 12,790 N/sq m. **131.** 300 litres. **132.** 119.5. **133.** 58. **134.** 72 carbon units in glucose, 48 carbon units in tartaric acid and 48 in propyl alcohol. **135.** 72 carbon units, 6 atoms. **136.** 7 atoms. **137.** 3 atoms. **144.** 69.72. **146.** 24.32. **147.** 87.62. **148.** 58.7. **149.** At. wt. 114.75; valence 3. **150.** 55.85. **151.** 118.7. **152.** 31. **154.** N$_2$O. **155.** H$_4$CON$_2$. **160.** Na$_2$S$_2$O$_3$· 5H$_2$O. **161.** MgSiO$_3$. **163.** C$_{10}$H$_8$. **168.** C$_2$H$_6$S. **169.** C$_6$H$_{14}$. **170.** Si$_2$H$_6$.

171. C_2H_6O. 172. CH_4. 173. C_2N_2. 179. 7. 182. 5.07. 211. 7.2 g Mg, 29.4 g H_2SO_4. 212. Alkaline. 213. About 14 tons. 214. 50 g. 215. 40 g and 80 g. 216. 2.5 moles. 217. 8.4 litres. 218. 162.5 g Zn and 1,225 g acid. 219. 2.8 litres. 220. 375 g. 221. 33.6 litres. 222. About 364.5 g. 223. About 146 litres. 224. 146.25 g. 225. 56 litres. 226. 5.6 litres. 227. 11.2 cu m. 228. 2.25 cu m. 229. 5.25 cu m. 230. 10.7 g NH_4Cl. 231. 28.7 g AgCl. 232. Alkaline. 233. 18 moles N_2, 14 moles H_2. 234. 2 moles SO_2, 11 moles O_2. 235. 0.08 mole $Fe(OH)_3$ was formed. 236. About 34.3 g $CuSO_4$. 237. 292.5 g NaCl. 238. 212 g Na_2CO_3, 222 g $CaCl_2$. 239. About 173 g. 240. 54 g. 241. 94.6%. 242. 4.9%. 243. 81.25%. 244. 5.85%. 245. 2.34 atm or 237,100 N/sq m. 246. 22.8 kcal or 9.53 kJ. 247. 83.4 kcal/mole or 3,492 kJ/mole. 248. 393.3 kcal/mole or 1,644.4 kJ/mole. 249. 14.45 kcal or 60.3 kJ. 250. 1.97 kcal or 8.35 kJ. 251. About 3,000 kcal or 12,540 kJ. 252. 110 kcal or 460 kJ. 253. 2.3 kcal/mole or 9.61 kJ/mole. 254. 4.8 kcal/mole or 20 kJ/mole. 255. 9.8 kcal or 41 kJ. 256. 57 kcal/mole or 238.3 kJ/mole. 257. 37.5 kcal/mole or 157 kJ/mole. 258. —27.6 kcal/mole or —115.4 kJ/mole. 259. 1,505 kcal or 629 kJ. 260.—12.5 kcal/mole or —52.25 kJ/mole. 261. 313.5 kcal or 1,310.4 kJ. 262. 7.6 kcal or 31.77 kJ. 263. 3.08 kcal or 12.8 kJ. 264. 283.5 kcal/mole or 1,185 kJ/mole. 265. 76 kcal/mole or 3,177 kJ/mole. 266. 5.2 times as great in the second case. 267. —116 kcal or —484. 9 kJ; —21.8 kcal or —91.1 kJ; + 7.6 kcal or 31.77 kJ. 287. 67.2 atm or 6,809,040 N/sq m. 288. 0.045 mole/litre. 289. 0.009 mole/litre. 290. [A]=0.24 mole/litre, [B]=0.6 mole/litre. 292. Increases 9 times. 293. Increases 12 times. 294. Increases 64 times. 295. v_1=0.03; v_2=0.0072. 296. (a) at equal rates; (b) twice as great in the second vessel. 297. At 200°C in about 0.16 min; at 80°C in 162.5 hours. 298. $[I_2]$=0.5 mole/litre; $[H_2]$=0.7 mole/litre. 299. $[N_2]$=5 moles/litre; $[H_2]$=15 moles/litre. 300. K=1.92; $[NO_2]$= =0.3 mole/litre. 301. 0.75 atm=75,990 N/sq m. 302. Reduced 1/20 initial pressure. 303. 50%; 83.3%; 90.9%. 304. No. 305. 0.45. 306. 83.3%. 307. 1.55 moles H_2. 308. $(CO_2) : (O_2)$=9 : 1. 309. [CO]=0.04 mole/litre; $[CO_2]$=0.02 mole/litre. 313. v_1 increases 8 times, v_2 increases 4 times. 314. In the first reaction both rates will increase equally. In the second: v_1 increases 8 times, and v_2 4 times. 315. v_1 increases 16 times, v_2 increases 4 times. 320. 12.5%. 321.189.1 g borax. 322. 430 g. 323. 5.1 g.324. 161 g. 325. 6.625 g.326. 342 g. 327. 12%. 328. 2,488 ml. 329. 125 ml. 330. 1,900 ml. 331. 62.5 ml should be diluted to one litre. 332. 161.5 ml. 333. 30.25%, 7.56 g-equiv/litre. 334. 11.8N. 335. 183.1 ml. 336. 4 : 15. 337. 250 ml. 338. 10 g. 339. 10 g. 340. 0.25N. 341. 0.3N, 24 ml. 342. 170 g. 343. 122.5 g. 344. 45. 345. 40. 346. About 27%. 347. 125 g. 348. About 1.4 litres. 349. 17.5 g. 350. 47.5 g. 351. 45°C. 352. 30°C. 354. 8.55 litres. 355. 33.2%. 356. 7.5 atm or 759,940 N/sq m. 357. 34.85% O_2. 358. 90.2% N_2O; 9.8% NO. 359. 10.1 kcal/mole or 42.3 kJ/mole. 360. 18.1 kcal/mole or 75.8 kJ/mole. 361. About 9°C. 362. 18.6 kcal/mole or 77.9 kJ/mole. 363. —9.4°C. 364. 18 kcal/mole or 75.36 kJ/mole. 365. 81°C. 367. 1 : 6. 368. 1.75 atm or 177,320 N/sq m. 370. 1.26 atm or 127,670 N/sq m. 371. 0.8 atm or 81,600 N/sq m. 374. 0.044 mole. 375. 18 g. 376. 30. 377. 342. 378. 186 mm Hg or 24,800 N/sq m. 379. 734.7 mm Hg or 97,950 N/sq m. 380. 0.41 mm Hg or 54.66 N/sq m. 381. 52.6 g. 382. 60. 383. 93.1. 386. 0.26°C. 387. About 102°C. 388. —27°C. 389. 1.62°C. 390. 18.4 g; 65.8 g. 391. About 2 : 1. 392. About —10°C. 393. 30. 394. 60. 395. 119.5. 396. 152. 397. 1.16°C. 398. 5.2°C. 399. 128. 400. S_8. 401. $C_6H_6O_4$. 405. 7.22×10²¹. 411. 1.86.

412. 2.48. **413.** 62%. **414.** 70%. **415.** 39%. **416.** 90%. **417.** 48.4%. **418.** 70%. **419.** 75%. **420.** 75%. **421.** 1.06. **422.** 4%. **423.** 2.16%. **424.** 4.032 atm or 408,540 N/sq m. **425.** —6.3°C. **426.** 89.68 mm Hg or 11,960 N/sq m. **427.** —7.4° C. **428.** Into three ions. **429.** 1.8 moles. **430.** 6.923×10^{19}. **431.** $[K^+] = 0.2$ g-ion/litre, $[SO_4^{2-}] = 0.1$ g-ion/litre. **432.** 0.18 g-ion/litre. **433.** $[Fe^{3+}] = 0.065$ g-ion/litre; $[Cl^-] = 0.195$ g-ion/litre. **434.** $[Ca^{2+}] = 0.18$ g-ion/litre or 7.2 g/litre; $[Cl^-] = 0.36$ g-ion/litre or 12.78 g/litre. **435.** 0.35 mole/litre. **436.** 0.125 mole/litre. **437.** 0.39 g-ion/litre. **438.** $0.1 N$. **439.** $[Mg^{2+}] = 1.32$ g-ion/litre; $[Cl^-] = 2.64$ g-ion/litre. **440.** 0.02 mole NaCl, 0.64 mole KCl and 0.24 mole Na_2SO_4 were dissolved. **442.** 5.5%. **443.** 0.045%. **444.** 2×10^{-4}. **445.** 3×10^{-7} **447.** 0.01 mole/litre. **448.** 2×10^{-5} mole/litre. **449.** 900 ml. **450.** 4×10^{-3} g-ion/litre. **451.** 1.85×10^{-2} g-ion/litre. **452.** About 42 times. **453.** 1.8×10^{-4} g-ion/litre. **454.** $[Pb^{2+}] = 1.5 \times \times 10^{-3}$ g-ion/litre. **455.** 1.7×10^{-8}. **456.** 7.9×10^{-5}. **457.** 4×10^{-12}. **458.** 3.2×10^{-9}. **459.** 1.35×10^{-8}. **460.** 0.013 g. **461.** 6.5×10^{-5} g. **462.** 2.6×10^{-2} mole/litre; 8.1 g/litre. **463.** About 117 litres. **464.** 4.5×10^{-2}. **465.** No. **466.** About 2,750 times. **467.** No. **468.** Yes. **484.** 2.7. **485.** 3.38. **486.** 11.4. **487.** 10.48. **488.** $[H^+] = 6.3 \times 10^{-7}$ g-ion/litre; $[OH^-] = 1.6 \times 10^{-8}$ g-ion/litre. **489.** $[H^+] = 0.425$ g-ion/litre; pH = 0.37. **490.** 6.02×10^{-7} ion. **491.** Reduced 10 times. **492.** Will be increased 4 units. **493.** 50 ml. **494.** 5.7. **546.** About 4 hours. **547.** About 7.2 days. **548.** 3.04 minutes; 4.38 minutes. **549.** 0.0003 g. **550.** 2,413 years. **551.** 1.95×10^{-3} mg. **553.** 4,740 years. **555.** 2 alpha and 2 beta particles. **556.** $K \rightarrow Ca$; $Rb \rightarrow Sr$. **557.** VI group. **558.** IV group. **559.** 134.4 litres. **560.** 206 g Pb; 112 litres of He. **565.** 77% Cl^{35}. **566.** 64% Ga^{69}. **567.** 85.54. **568.** 63.54. **577.** Reducer. **594.** 16 g. **595.** 5.44 g. **599.** Emf is 1.93 V. **602.** Emf is 0.8 V. **603.** Only Zn dissolves. **605.** Emf is 0.46 V. **606.** Emf is 0.61 V. **607.** Emf is 2.52 V. **608.** 0.41 V. **609.** Will not change. **610.** Hydrogen electrode. **621.** 63.6. **624.** In the first solution. **627.** 0.58 g. **628.** 2.41 A. **629.** 1.34 A. **630.** 10 hours, 43 minutes, and 20 seconds. **631.** 2 minutes and 5 seconds. **632.** 5.6 g. **633.** 894 coulombs. **634.** 9.8 g. **635.** Will be reduced 2.74 g. **636.** About 5.6 g Cd; 0.56 litre O_2. **637.** 1.94 g. **638.** 2.7 g $Mg(OH)_2$; 3.3 g Cl_2. **639.** 2,412,500 coulombs. **640.** About 7.1 g of I_2, 0.627 litre of H_2. **641.** 48.81. **642.** 69.67. **643.** 56.2. **644.** 114.72. **645.** 0.329 mg. **646.** 55.84. **647.** 24.5 g. **648.** 1.118 mg. **650.** 46.2 g. **651.** 72% Ag. **652.** 750 g. **653.** About 61 g. **654.** Mg_3Sb_2. **655.** $MgCu_2$ and MgCu. **656.** 617.3 g. **661.** Aluminium. **663.** 525 kg CaH_2; 1,625 kg Zn; 2,450 kg H_2SO_4. **665.** (a) 1.2 tons, (b) 111 tons. **666.** 0.9 g; D about 0.7. **675.** 61.25 g $KClO_3$. **680.** 134.4 litres. **683.** 1,000 ml O_3; 1.52 kcal or 6.36 kJ; 1,500 ml O_2. **686.** About 2%. **690.** 28.57 g S. **692.** 13.44 litres of SO_2. **697.** 48.76 g; 24.38 g. **698.** Equal quantities of the acid are spent for dissolution of nickel in both cases. The required quantity of the concentrated acid is 66.45 g and of diluted acid is 132.9 g. **704.** 146.4 litres. **708.** 67.2 litres. **727.** 79.08 g P; 395.4 g $Ca_3(PO_4)_2$. **741.** 215 litres. **749.** 36.9%; 28 litres; 56 litres. **752.** The thermal effect of the first reaction is 52.8 kcal or 221 kJ; in the second reaction it is 31.4 kcal or 1,313 kJ. **753.** 5,400 kcal or 22,600 kJ. **770.** 132.5 g. **771.** 20.6 kcal or 86.11 kJ. **774.** 4 hours, 11 minutes, and 16 seconds. **778.** 50%. **781.** 285.7 kcal or 1,194.2 kJ. **784.** 12.5%. **798.** 3 mg-equiv/litre. **800.** 105.8 g. **801.** 5 mg-equiv/litre. **802.** 144.6 mg. **803.** 2 mg-equiv/litre. **804.** 2.14 mg-equiv/litre. **805.** 9.1 mg-equiv/litre. **812.** 0.8 g. **830.** 294.6 g. **843.** 2.63 g; 1.58 g. **846.** —5.4 kcal or —22.6 kJ. **847.** 5140° C. **850.** 0.3 mole.